SHOCKING
THE SUBURBS

Jago Dodson abandoned a promising early career in the petroleum sector to study anthropology and urban planning. Since finishing a PhD at Melbourne University in the early 2000s, his research work has investigated the multiple socio-economic linkages between housing, transport, government and urban planning in Australia's cities. He has been a regular contributor to public debates around transport and housing in Australia through his work at Griffith University's Urban Research Program and he has consulted to governments on urban matters.

Neil Sipe moved to Australia upon completing his PhD at Florida State University. He has taught in the Griffith School of Environment since 1998 and has served as head of the urban and environmental planning program from 2002 to 2006 and again from 2008 to the present. Before coming to Australia, he was a planning consultant in private practice for 10 years. He has also worked in the investment banking industry as a financial analyst and as an economic analyst in the University of Florida Business School.

Briefings

A series of topical books exploring social, political and cultural
issues in contemporary Australia

Series editors: Peter Browne and Julian Thomas
Australian Policy Online (www.apo.org.au)
Institute for Social Research, Swinburne University of Technology

SHOCKING THE SUBURBS

oil vulnerability in the Australian city

JAGO DODSON and NEIL SIPE

UNSW
PRESS

A UNSW Press book

Published by
University of New South Wales Press Ltd
University of New South Wales
Sydney NSW 2052
AUSTRALIA
www.unswpress.com.au

National Library of Australia
Cataloguing-in-Publication entry

Author: Dodson, Jago.
Title: Shocking the suburbs: oil vulnerability in the Australian city/Jago Dodson, Neil Sipe.
ISBN: 978 1 921410 32 1 (pbk.)
Subjects: Urban transportation – Australia.
     Automobiles – Australia – Fuel consumption.
     Petroleum products – Prices – Australia.
Other Authors/Contributors: Sipe, Neil G.
Dewey Number: 388.40994

Design Josephine Pajor-Markus
Cover iStockphoto
Printer Ligare

This book is printed on paper using fibre supplied from plantation or sustainably managed forests.

Contents

1 An unfamiliar world   7

2 Solving housing problems by creating suburbs   11

3 Towards uneasy oil   21

4 Shocking the city: Higher fuel and housing costs   37
   in the suburbs

5 Oil and politics: The impact of rising fuel costs   53
   on suburban constituencies

6 Oil-proofing Australian cities: Transport and   67
   planning solutions to oil vulnerability

7 Aftershocks: Facing the challenge of oil   86
   vulnerability in Australia's cities

   Acknowledgments   89

   Notes   90

CHAPTER 1

An unfamiliar world

At 10:15 on a Tuesday night a long queue has formed outside a petrol station on one of Melbourne's outer suburban traffic arterials. Grim motorists clutching supermarket dockets look wearily at the two lines of numbers ticking over on the pump. The upper line is fast outpacing its neighbour below. One of the drivers looks anxiously up at the forecourt sign reading '169.9 cents per litre' and calculates the hit he'll take as he fills his ageing 6-cylinder. Inside, near the cash register, newspapers tell tales of angry and despondent motorists wondering how to maintain their suburban lifestyles and livelihoods as the cost of fuel inflates mortgage interest rates. In Australia's Parliament, blame and invective fly more furiously with each upward tick of the bowser price. It's getting harder to keep up with the daily shifts and contortions in the global energy supply. Almost all the petroleum portents signal even higher fuel prices. For those wincing at the suburban bowser the stress will only mount.

Strange, unfamiliar and unsettling events are afoot around our world. In Mexico, production from Cantarell – the world's fifth-largest oil field – is plummeting by 13 per cent each year. Oil prices have increased from just US$25 per barrel in early 2004 to over $140 by mid 2008. In March 2008, tens of thousands of Buddhist monks rioted over the Myanmar government's cuts to fuel subsidies. In the United States, food wholesalers reported runs on rice as global food prices shot suddenly higher. Some US restaurants have had their used cooking oil stolen for use as biodiesel. Saudi Arabia's oil engineers are struggling to squeeze another few hundred thousand barrels per day from the kingdom's massive oil resources after US President George Bush begged Saudi King Abdullah to increase supplies. Barely a year earlier the United States established a new military command – AFRICOM – as the attention of international petroleum companies turned to new unexplored oil territories, such as Sudan, Western Sahara, Chad and Equatorial Guinea. Russia is newly ascendant and basking in wealth as it becomes the world's second-biggest fuel pump. And in the Persian Gulf in early 2008, Iranian gunboats duelled with US warships as tense military commanders anxiously fingered their panic buttons. Oil prices shot up to a record $138 per barrel in May 2008 after Israel threatened to attack Iran over its nuclear activities. In Asia, millions of newly prosperous Chinese and Indian citizens are discovering the joys of motoring as their countries' growth spurs new energy cravings. In the United States, General Motors has announced that it will dump the iconic gas-guzzling 'Hummer' SUV as gasoline prices make it almost unsellable. European truck drivers are blockading highways over higher petrol prices while world leaders are calling for concerted action to cut the price of oil.

The early months of 2008 have offered us a sharp glimpse of the strange and unfamiliar world we're facing. By mid 2008 some commentators were speculating that the oil price could pass $200 per barrel by the end of the year – equivalent to AUD$2.00/L for petrol – though others thought the oil price could fall hard. The dramatic price rises and wild uncertainty in oil markets have already made the cheap and easy oil of the past few decades a distant memory. Strains are showing in global petroleum supplies and this is bringing powerful and unwelcome forces to our cities and suburbs. Already frictions are apparent in our suburban transport and housing systems as oil and petrol become more expensive. The costs of driving have increased sharply, and this has permeated through our economy, pushing up mortgage interest rates and food prices.

Since World War II, the suburbs of Australian cities have been built on the premise of cheap petroleum. For all their flaws, Australia's suburbs have proved astonishingly successful, and have offered millions a decent living untroubled by the ills that affect cities elsewhere. As our colleague Brendan Gleeson has argued, the suburbs are Australia's 'heartlands'. But their dependence on cheap fuel has been ignored for far too long by residents and those who represent them. This neglect must end if the suburbs are to sustain and nourish Australia's national life in the coming decades.

Many households are already deeply worried about the cost of fuel. Some have drastically changed their lifestyles. Those with large housing debt fear the possibility of default, or worse, repossession. The outer suburbs, where the less wealthy, the more car dependent and the most indebted live, are worst affected – even if they are perhaps also the most hopeful.

The strongest calls for relief will sound from these places as the global petroleum challenge looms larger. Australia's political representatives will be summoned to protect the suburbs from this threat. Suburban 'battlers' were transformed into 'aspirationals' in the early part of the decade, but under petrol and mortgage pressure they have morphed back into 'working families'. Their demands for security and salvation will press firmly on Australia's political levers. Reconstructing the suburbs so that they continue to sustain all our citizens – and secure their votes – is a new challenge for our policy makers.

Meeting the energy challenges of our suburbs in the face of costly crude will be neither simple nor easy. The days of cheap oil are gone, probably forever. We can't simply sit back and hope that 'markets will save us' or that something magical will come along. Our political leaders and policy shapers will need to be more active in organising and managing our cities and suburbs. There is little evidence that this has started. Most of Australia's current metropolitan plans ignore petroleum issues, and there are few signs that our planners and elected representatives realise that serious government intervention will be needed to respond to declining energy security.

Any such response will also bring great risks. The main challenge is to avoid blind faith in anything. While some magic fuels and wonder vehicles that are as convenient as the motor car may appear, there is considerable doubt that they will be able to be run for the same price as conventional petroleum. If Australia's suburbs are to be secured from the economic and social threats of higher fuel prices we must do more; we must reconstruct our suburban transport systems to offer residents viable and affordable choices in the era of expensive energy. Our state governments, with federal help, must redress their failure to extend our public transport infrastructure as our cities have grown. Most importantly, we urgently need a program to revitalise the way we plan our suburban public transport.

We offer this book as a piece of public writing that engages with the risks our cities and suburbs face, the effects they are already feeling and the problems we must grapple with if we are to secure our suburbs' future. We focus on the relationship between housing and transport that formed and structured Australia's suburbs and on today's dramatic and unsettling shifts in this relationship and the structures it supports.

This writing draws on our research at Griffith University's Urban Research Program where, since 2005, we have been investigating and evaluating the effects of rising fuel prices and higher interest rates. Our intention, as urban planning researchers, is to advocate for new thinking to address the problems looming in our cities and suburbs and to push for new policies and programs that can soften the jolts to our suburban transport and housing systems as they travel over the rough new landscape of energy risk. We also recognise that Australia's oil-dependent suburbs house its most vulnerable urban residents. The choices facing many of our suburban citizens are already heavily constrained. Higher fuel prices will only add – unfairly, inequitably and disproportionately – to these difficulties. We argue that Australia must act now to ensure that the world of constrained energy supplies does not deliver painful and unremitting shocks to our cities and suburbs.

CHAPTER 2

Solving housing problems by creating suburbs

The vulnerability of Australian cities to oil and mortgage debt pressures has its origins in the historical conditions of their development since the mid 19th century. While this is not the place for a long history lesson, we ought to comprehend the social and economic shaping of our cities, especially how their housing and transport arrangements developed.

In this chapter we venture on a short historical tour through the emergence and development of Australian cities, focusing on two essential factors that shaped their current vulnerability. The first is the reliance on suburban home ownership as a method of providing housing to Australia's population. The 'dream' of home ownership has been tenderly nurtured in our suburban nurseries.

The second is the heavy reliance on motor vehicles for urban transport. The expansion of home ownership beyond the reserve of the privileged and wealthy was achieved on the back of urban transport networks – first by tram and train and then by automobile, arterial road and freeway. The car is now overwhelmingly the suburban travel mode.

Government is deeply implicated in moulding and fostering Australia's urban matrix of housing and transport, despite often disavowing this role. And when the problems of housing debt and rising fuel costs become acute, it will be to government that our

urban populace turns for resolution and salvation. Our governments must be ready with responses to these challenges to preserve
and secure Australian cities.

The rise of the Australian city

Australia's major cities, for all their faults, function quite well as
places for living, and often appear near the top of international
scorecards of 'liveability'. They lack the crime, the environmental
damage, the social dysfunction, the seething population pressures
or the economic woes that afflict many cities around the globe.
Urban Australians are typically well housed – most dwellings are
of good quality on spacious lots, and transport networks allow for
most urban residents to travel easily across their realm.

A glance at the history of Australia's largest cities, however, shows
that they were not always so. The Australian city of the late 19th
century could be an unpleasant and unappealing place. Most industry was situated close to commercial centres or ports and sourced
labour from a workforce that was housed nearby – the labouring
classes dwelt within walking distance of their workplaces. High
land values pushed up residential densities, creating crowded and
unhygienic living conditions – as late as 1901 Sydney suffered an
outbreak of bubonic plague. One historian described 19th century
urban housing thus:

> They had been thrown together in the skimpiest way using the
> poorest materials, their inadequacies hidden under a smother
> of plaster. Fifty years later they were damp, ill-lit, unhealthy
> hovels infested with rats and disease, and were a breeding
> ground for vice. They had become dilapidated, rickety piles
> of rubbish, their wavy walls shored up with baulks of timber
> propped across narrow alleys and with sagging ceilings, rotted
> woodwork and peeling plaster.[1]

Escape from the slums

Australian governments dabbled in regulating urban housing quality through rules controlling lot size, building design and boundary
setbacks. But they did little to support social access to housing. The
middle and upper classes could afford more than a squalid worker's
hovel, and many left the city to find refuge in gracious abodes in
the surrounding countryside. But this created a new tedium in the
daily return to the city for work, business and leisure. One author
writing on 19th century slums observed that 'without cheap, rapid
and conveniently scheduled transit the suburb could not offer a way
out for the slum dweller'.[2] The horse-drawn omnibus had appeared
in Australian cities by the mid 1800s; trams and railways were in
extensive use by the century's end. They enabled those wealthier
citizens to live far beyond walking distance from their places of
work. Thus was the modern suburb founded, marking a new set of
urban social arrangements. By taking people away from the foetid
effluence and sooted industrial atmosphere of the city, the trams
and trains helped place the suburban realm as the land of the bourgeois, flushed by refreshing air and health-giving light.

Australian suburbs became more than this, though. Workers
who could afford suburban life found that the gardens and yards
of their new homes also offered economic opportunities: for petty
livestock husbandry, food growing and the frugal fabrication and
fixing of all forms of household goods, from furniture to fowl-
coops. The capacity of the Australian suburb to foster resilience
among the working class has long been recognised as among its
greatest strengths.[3]

On the rails to own your own

Suburbanisation depended on a dynamic land market, and land
booms were prominent features of the 19th century Australian urban economy. Melbourne suffered from extreme bouts of
property boom and bust during the late 19th century. But land
speculation depended on buyers who could access the proliferating

far-flung plots. Urban railway systems expanded quickly in Australian cities as governments and private speculators funnelled capital into suburban rail lines, through such measures as Melbourne's 1894 'Octopus' Act. Funds invested in rail brought huge returns through high lot prices around the train stations. New villages with vibrant commercial hubs sprouted around the suburban stations. Some contemporary observers derided this new urban pattern as an ungainly Australian 'sprawl'.[4] Yet this was a concentrated sprawl – as the radial lines stretched to the horizon, the new suburban settlements could spread little further than walking distance from their station. Beyond that radius from the terminus, suburbanites became pedestrians and land prices fell sharply, leaving a semi-rural landscape of green wedges between the rail lines. Motor bus services began servicing those areas by the 1920s but they were run by private operators, and remained undercapitalised and underdeveloped.

Though the Australian 'dream' of 'owning your own' had emerged by the late 19th century, in the second decade of the 20th century most households in Australian cities still rented from private landlords. In Sydney, for instance, home ownership grew only from 28 per cent of households in 1891 to 40 per cent by 1921.[5] A new geographic divide had also emerged: between the renters in the city's core and the owner-occupier suburbanites.

Home ownership achieved political resonance in Australia in the early 20th century. Australian governments were attempting to solve the problem of the slums, and thus promoted suburban home owning via Workers Dwellings Acts or Savings Bank and War Service Home schemes. Some countries began experimenting with urban public or socialised housing schemes and multi-unit housing, but in Australia, mass suburban home ownership became a bipartisan aspiration.

Despite strong government promotion, it was not until 1947 that home owners attained majority status in Australia. By then the country had seen 18 years of either war or depression, both of which had slowed the suburban surge. The population was growing quickly, and faced acute housing shortages. There was widespread overcrowding in the cities, compounded by rapid immigration. Housing formed the core of Australian governments' postwar social 'reconstruction' programs. The Commonwealth funded 36 per cent of all Australian housing constructed between 1945 and 1970,[6] while state housing authorities drew on that finance to build 14 per cent of Australia's housing stock during that period.[7] From the mid 1950s until the 1970s, state housing authorities built extensive outer-suburban estates which they on-sold to home owners.

As the economy grew, building societies and savings banks also provided a growing pool of housing finance that stoked the new economic engine of suburbia. The value of Australian mortgages grew from just £100 million in 1946 to £682 million a decade later.[8] The result of this frenetic postwar housing activity, especially the expansion of credit to the masses, was a massive increase in the home ownership rate during a very short period. Just 53 per cent of Australian households owned their homes in 1947, but 72 per cent had achieved this feat by 1965.[9] Within less than 20 years, Australia had become a mass home-owning society. At end of the 1960s, 70 per cent of households nationally owned outright or were purchasing their home; for some cities (such as Brisbane), it was 75 per cent. As new home owners reached for their credit-fuelled 'dream', the suburbs spread further.

Driving the suburbs

The postwar home ownership boom unleashed a massive housing construction surge that washed a new wave of suburbanisation across Australia's cities. Population levels fell near the central business district but climbed dramatically in areas further out.[10] In Sydney, the inner and central city share of the metropolitan population fell from 38 per cent to 25 per cent between 1947 and 1961, while new fringe zones expanded their share from 44 per cent to 62 per cent. Some outer settlements, such as Blacktown, more than doubled their residents in a few years during the 1960s.[11] The concentration of debt-laden home purchasers among the new subdivisions and estates ringing the city edges soon saw these zones termed 'mortgage belts'.

Like earlier booms, the postwar suburban rush was enabled by rapid subdivision of fringe land and weak and uncoordinated government planning. Self-building was common: it often began with a shack that was then extended as finances allowed. The modest finances of the new suburbanites also meant meagre local government infrastructure budgets. Poor integration, coordination and control of development meant that state infrastructure lagged far behind suburban subdivision. A quarter of the roads in Australian capital cities remained unsealed as late as 1969.[12] The Whitlam Labor government made some headway in redressing suburban infrastructure deficits during the 1970s, but many postwar subdivisions relied on septic tanks for their effluent disposal until the 1980s, and even in the 1990s, dirt roads could be found in some parts of Australian cities.

Unlike the earlier phases of Australian suburbanisation, the postwar boom led to people travelling in cars, not trams and trains. The reasons for cars becoming the suburban vehicular norm are many. At the household level, rising incomes meant greater purchasing power and thus greater consumption. Car ownership rose quickly on the back of the suburban home-owning rush. While there were only 80 cars for every 1000 people in 1950, this rate more than doubled by 1960, and had reached 330 by 1973.[13] After the purchase price, the costs of running a car were relatively marginal, especially when fuel rationing ended in 1950.

But it would be a mistake to view the rise of the car-dependent suburbanite as a triumph of consumer choice. Government policy played a major, if not directive, role. The Commonwealth encouraged a domestic motor industry – it would be a symbol of Australia's industrial vitality. Public transport networks had become run down, as a result of high demand and low investment during the Depression and World War II. The new outer suburbs had not been provided with extended tram services or new rail infrastructure. Instead, grandiose freeway schemes, from an even more rapidly suburbanising and prosperous United States, were inveigling their way into state government metropolitan plans. Freeways gnawed into transport budgets, leaving little for new public transport links. Trams and cars came into conflict in crowded city streets. Perth and Adelaide removed their trams in 1958; Sydney and Brisbane did the same in the 1960s.

From the 1950s on, the dominance of the car was entrenched by successive urban plans. Sydney's 1948 County of Cumberland Plan was the first of these schemes and it marked out routes for inner-city 'distributors' and suburban expressways. In the 1960s, Australian governments hired American road engineers to draw up extensive freeway blueprints inspired by Los Angeles and Detroit. Brisbane's 1965 freeway plan turned the north bank of its river into a slow sludge of traffic that extended out through its southern suburbs. Melbourne's road building has closely followed the 1969 freeway plan that was the origin of the South Eastern and Tullamarine freeways, City Link and the Western Ring Road. While governments claimed merely to be 'predicting and providing' for the new travel choices of their suburban populace, their policy decisions in fact made such 'choices' inevitable.

The city strikes back

Australia was fortunate that its freeway suburbanisation program began a decade later than the equivalents in American cities: we avoided the 'doughnut' effect, where prosperous suburbs encircle an abandoned and impoverished city core. Weakening suburban industry and inner-urban gentrification remade the social geography of Australia's cities. The growing 'information' sector meant new jobs in the city centres and stimulated a reversal of the suburban rush. The retaking of the inner city by the bourgeois youth brigades began pushing up central-city house prices from the 1970s; they increased faster than prices in the suburbs. The remnant working class, including large numbers of migrants, were often displaced, exiled to outer suburban zones. Unlike in the United States, where impoverished city cores are the norm, Australia's poor mostly huddle on our cities' edges.

Hundreds of thousands of suburban manufacturing workers lost their jobs to restructuring between 1971 and 1992. Many prosperous middle-suburban communities subsided into uncertainty and disadvantage, creating an impoverished 'third city' between the new fringe master-planned estates and the city core.[14] Desperate state governments chased international capital for the city centres, which were refashioned as sites of conspicuous consumption and wealth; meanwhile, many outer suburbs faced growing socio-economic vulnerability. In some suburbs, the sense of insecurity translated into electoral volatility as Labor and Liberal both courted the struggling post-industrial 'battler'.

The suffering in the suburbs and the 'flexibility' of the information worker unsettled the Australian 'dream'. Home ownership levels peaked at 71.4 per cent in the late 1960s and gradually slipped to 66.2 per cent by 2001. In part this was due to strong growth in higher density development from the late 1960s. The majority of people in detached suburban houses are home owners; the majority of those in higher density development are renters. The higher land values and stronger rental market in city centres has concentrated higher density development there, increasing the contrast between the city and suburbia.

During the 1980s and 1990s, government let markets loose among the cities. Unlike the direct economic intervention in housing of the postwar decades, the Commonwealth left investment to market forces (albeit while continuing to plan freeways) and modulated national activity through interest rate settings that responded only to inflation. Governments avoided direct housing construction and focused on managing what was left of public housing, doling out Rent Assistance to pensioners and cash to new home owners.

Financial deregulation, followed by a cycle of low interest rates in the late 1990s, encouraged investment and unleashed a new housing boom across Australian cities. House prices in Australia's cities more than doubled as credit-laden households indulged in a frenzy of auction and speculation. Greater maturity among funding institutions and greater capitalisation of the development sector saw semi-exclusive master-planned estates – boasting infrastructure such as lakes, bike paths and, for some, gates – pockmark the suburban fringe. The diversifying suburban service economy threw up higher income households who settled among these new promised lands of owner-occupation. The 'McMansion' rose from the suburban plain as a symbol of this land and credit splurge.

This period also saw new suburban freeways. Governments in most states found new friends in private finance to fund freeways, further embedding the car as the essential mode of suburban transport. And the Commonwealth maintained a significant urban road funding role through its Auslink program. Many of the new roads have allowed private 'roadlord' companies to extract tolls from suburban motorists: Sydney residents, for example, have seven toll road systems in their city. Public transport investment has languished. Development of new suburbs is rarely integrated with transport. In Melbourne, new suburbs at Epping and South Morang have been denied their promised rail links. Brisbane's new Springfield satellite town will be substantially completed before its rail link arrives in 2015. In Sydney the state government continues to equivocate over the northwest line. Given what we know about our petroleum future, these 'ghost' trains may be running disastrously late.

Getting here from there

Automobiles now account for around 75 per cent of trips in Australian cities and this level is increasing. Car use in southeast Queensland grew from 78 per cent of travel in 1992 to 80 per cent in

2004;[15] more than 81 per cent of all trips in Perth are by car.[16] While 73 per cent of trips in Melbourne are taken by car,[17] even in Sydney, which has the highest public transport patronage of any Australian city, the figure is 70 per cent.[18] The heavy dependence of Australian cities on cars for transport equals a heavy dependence on petroleum. Our suburban armies march on their petrol tanks. Australia's suburban residents are among the most car dependent outside the United States. As we show in the next chapter, the long-running infrastructure deficits in our suburbs have now placed them in a vulnerable position.

The shocks to the cost of suburban transport raise questions about the sustainability of our suburbs, and our debt-ridden system of housing can only add to our fears. The housing boom of the late 1990s pumped up housing prices and loaded many suburbanites with massive debt. The average mortgage grew by 125 per cent in Australia between 1994 and 2004, from $91,200 to $205,600. As interest rates have climbed in the late 2000s, on the back of inflation caused by rising fuel prices, the cost of mortgages has further increased. By early 2008 mortgage holders had faced their eighth interest rate increase since 2004 – the media has made considerable hay out of anguished suburban home owners. As the price of petrol continues to rise, inflation will further tighten the mortgage belt. Could higher fuel prices mean high water for the Australian suburban tide? What is the nature of the oil threat that menaces our city? It is to the great uncertainty posed by that incredible suburban lubricant, oil, that we now turn.

CHAPTER 3
Towards uneasy oil

The age of easy oil is over.[1]

When we fill our cars with petrol we rarely stop to think about the pungent liquid pouring from the pump. We guzzle 3.6 billion litres of petrol each day – 1300 billion litres every year.[2] Petrol, like other liquid fossil fuels such as diesel or aviation kerosene, is refined from crude oil. The world consumes slightly more than 86 million barrels of crude oil every day. The cost of petrol is, of course, tied to the price of oil. The past four years have seen a dramatic increase in the global price of oil. During the late 1990s the oil price stayed at around US$10, but it had doubled to around $20 by the early 2000s. Since then it has raced to over $130, growing 40 per cent on average during each year of the past half-decade. In March 2008 global oil prices passed their record inflation-adjusted high of $104 per barrel. This price growth has been in marked contrast to the stable patterns of the past two decades. The current high real price of petroleum is unmatched by any other period in the past 50 years, even the oil shocks of 1973 and 1979.

There has been voluminous commentary about the causes of the rising oil price since 2004. Explanations have ranged from rising geopolitical tensions to oil production constraints and supply threats mixed with mismatches between rapidly expanding global demand and world production capacity. More ominously, it has been suggested that the world's petroleum supply may be close to a production peak, and oil reserves will soon begin to deplete. Depletion of oil reserves implies a widening gap between the desires and expectations of international consumers and the ability of oil companies to meet these desires. A peak in global oil production would mean great uncertainty for cities and suburbs that were set up on the basis of infinite cheap petroleum. If we are to recognise and address the many challenges that likely future oil price hikes pose for our cities, we need to understand the global energy security situation.

More is more

Growing demand for petroleum has been among the biggest factors underpinning the recent hikes in fuel prices. Growing demand, from businesses and households, is the result of greater economic activity. World economic growth has galloped over the past two decades, from around 2 per cent annually in the late 1980s to around 5 per cent today. It hit highs of 5.3 per cent in 2004 and 5.4 per cent in 2006. This rapid growth has occurred most dramatically in China and India, which together account for 37 per cent of humanity.[3] The Chinese economy has grown more than 7.5 per cent each year since 1991, and has not fallen below 9 per cent since 2002.[4] India's GDP growth has remained above 4 per cent since 1991, and above 7 per cent since 2002.[5]

The incredible economic momentum of China and India is drawing millions of rural peasants to burgeoning mega cities where they become workers, consumers and petroleum users. Oil guzzling by developed nations was already putting pressure on petroleum prices. Chinese petroleum demand more than doubled between 1995 and 2005, from 3.4 million barrels per day to 7.4 million barrels per day.[6] The International Energy Agency (IEA) expects that China and India will together be responsible for 43 per cent of world oil consumption by 2030 and that their total petroleum use will more than double to 23.1 million barrels per day by 2030, up from 9.3 million barrels per day in 2005.

But India and China are not alone – other populous and growing nations, such as Brazil, Russia, Iran and Saudi Arabia, are also raising their oil demand. The global petroleum thirst has grown briskly over the past decade, from 69.5 million barrels per day in 1995[7] to 82.5 million barrels per day by 2004 and 86 million barrels per day in 2007.[8] The IEA predicts that by 2030, global oil demand will grow to 116 million barrels per day, a 37 per cent increase on 2006 levels. Unless the world can expand its oil supplies by more than a third, it will be unable to supply this amount, making much higher petrol prices inevitable. With demand hot on the heels of supply, any threats to the supply of oil can bring large price jolts as consumers rush to secure their share. There is also considerable anxiety about whether or not the world's largest oil producer, Saudi Arabia, will be able to continue to expand its production.[9]

Pump it up!

As well as rapidly growing global oil demand, capacity constraints in oil production – including old wells, rusting pipelines and ageing refineries – may add further risk to the security of petroleum supplies. The oil price jolts of the past few years show that oil producers haven't been able to boost their production at a sufficient rate to meet growth in demand. Given the massive expected demand growth, it's likely that this deficit will only widen unless new supplies can be brought on-stream.

Reports earlier this decade suggested that the global investment needed to meet petroleum demand growth by 2030 was approximately US$3 trillion, or $100 billion per year.[10] In 2006 the IEA estimated that $4.3 trillion[11] worth of oil wells, pipelines and refinery investment would be needed to satisfy expected the 2030 demand. A 2008 estimate suggests that $100 trillion is needed![12] To compound the problem, the infrastructure needed – oil exploration and drilling, pipelines, transfer facilities and refineries – involves huge and complex projects which take years, even decades, of planning. The world faces long time lags in the rollout of new oil production facilities, in the face of already rising prices. Worse, higher fuel prices further increase the capital cost of oil production, making such investment even more difficult to justify and secure. The task of assembling and coordinating the investment that is needed to meet future oil demand is viewed by many industry analysts as an almost insurmountable challenge, even before the question of the physical and geopolitical security of such investment is contemplated.

In all the wrong places

Adding to all this is a shifting geography of global oil reserves and production. Geopolitical instability is a factor in the current high price of oil because it increases the investment risks for exploration and production. The US Government Accountability Office has calculated that 36 per cent of global oil reserves are in 'high-risk' locations and a further 15 per cent are in 'medium-risk' places.

Oil reserves are slowly becoming concentrated in a shrinking number of Middle East nations. The Middle East is the most geo-politically significant oil-producing region: it currently provides 62 per cent of global oil production. Major Middle Eastern oil producers, such as Iran and Saudi Arabia, both of which are control-led by jittery authoritarian regimes, and which together hold a third of the world's oil, are barely able to manage their politically difficult mix of ever-increasing populations and active Islamist militancy. Both of those countries are using their massive oil revenues to quell their restive populations through social programs and subsidies for consumer goods. Iraq, which holds the second-largest national crude oil reserves – almost a tenth of the global total – is wracked by foreign occupation, insurgency and multiple ethno-political and religious tensions. Investment in new production facilities at the magnitude needed to supply future global oil needs is unlikely in such an economically and politically unstable region.[13]

Beyond the Middle East a further set of major oil-producing nations also face political insecurity. Nigeria, Venezuela and the former Soviet republics of Central Asia – together, this group holds

13.5 per cent of world oil reserves – all have to deal with a number of political and other internal stressors. In Nigeria, for instance, rebels have made frequent and disruptive attacks on Niger Delta oil installations. Venezuelan President Hugo Chavez has repeatedly threatened to cut off the 10 per cent of US oil imports that flow from his country. Russia has tightened control of its energy resources and shown a strengthening confidence in using its 6.6 per cent of world oil reserves to strategic advantage; it cut vital natural gas supplies to Europe in 2006.

Supply uncertainty due to political instability also encourages speculation in oil markets, which can cause volatile price swings. Investors face heightened 'risk premiums': investment costs that are raised to account for the chance that war, conflict and unrest, or political instability, could disrupt the flow of crude. Petroleum consumers are forced to set up complex financial 'hedging' arrangements to ensure stability in their petroleum costs. In this environment short-term disruptions to supply, such as insurgent attacks on oil pipelines or the threat of military clashes in production zones, can frighten speculative prices to levels far beyond those caused by simple production lags.

The threat of terrorism or piracy, especially along pipelines, at oil refineries or within strategic sea-lane 'choke points', is another security problem for petroleum markets.[14] Approximately 35 per cent of global oil deliveries pass through either the 9.5 km wide Straits of Hormuz sea lane in the Persian Gulf or the 2.5 km wide Straits of Malacca between Malaysia and Indonesia. Anything that prevented tanker passage through these sea lanes would send global oil prices upwards.

So the outlook for global petroleum supplies in terms of security remains shaky at best. The odds that international investors might spend $4 trillion to boost global oil production over the next 20 years seem long unless higher prices improve financial returns. But higher oil prices, and/or any medium-term shortfall in supply, could further heighten geopolitical tensions, especially if large oil-consuming nations seek to capture supplies directly. Faith in the ability of international oil markets to meet growing consumer demand could easily be broken by geopolitical energy competition as consuming nations manoeuvre for advantage or militant actions amplify market jitters. Beyond the immediate and near-future uncertainties generated by geopolitical risks, demand growth and supply constraint leave the longer term availability of petroleum shrouded in doubt.

Petroleum geopolitics could also have other severe global consequences. Ongoing international dialogue and cooperation on responses to climate change presume a high level of international stability and mutual trust. It is unlikely, for example, that an international climate treaty could be sustained through a long period of competition for access to petroleum supplies, especially if that competition escalated to outright hostility. The temptation to renege on climate promises in the face of tightening energy supplies could prove too great for many countries, especially those already prone to internal unrest. Clearly, there is a lot more at stake in global energy security than just crude oil.

A fit of peak?

If production constraints and supply jitters were not enough to jangle the nerves of petroleum producers and consumers, the notion of peak oil is certainly a stimulus to long-term paranoia. The dramatic oil price increases of the past four years can be easily explained: they are the result of rising demand running into supply constraints that have been amplified by geopolitical tensions. But such fears and rising prices have drawn attention to the fact that global oil reserves are finite. Many petroleum industry commentators predict that the world will hit a point at which total oil production reaches a maximum, a 'peak', and will then decline as global oil reserves are used up.

The theory of 'peak oil' was first proposed by Shell Oil geologist Marion K. Hubbert in the 1950s. Hubbert drew on historical production data from individual oil wells and aggregated this to the regional and the global scale. It was well known in the oil industry that when drilling commences in a new oil reservoir, the rate of outflow typically accelerates rapidly. Sinking new wells ramps up this production pace. Eventually, though, a point of maximum flow occurs, after which production rates decline no matter how much is done to boost output.[15] This pattern gives a broadly 'bell-shaped' curve over time: extraction accelerates initially and then plateaus before dropping over the peak and slipping away in an extended taper. The pattern is also known as a 'Hubbert' curve.[16]

Many oil regions have already seen their production peak and decline. The United States peaked in 1972 and the North Sea (United Kingdom and Norway) rolled over in 1999. Australian oil production peaked at just under 214 million barrels in 2000 and had slipped to 124 million barrels by 2006.[17] In 2004 the Australian government estimated that there is a 90 per cent chance that Australian crude oil production will decline to just 72,000 barrels per day by 2020, even taking into account 'undiscovered' reserves.[18] Australian consumption, however, is expected to hit 800,000 barrels per day as early as 2009. This will create massive post-peak dependence on foreign oil.

By combining and projecting data from individual oil fields and regions into a single global curve, the point of global 'peak' oil production can be estimated. Most projections of economic prosperity assume continuously increasing world consumption of petroleum over the coming decades. A peak in global oil output, followed by a decline, would mean an eventual shortfall between global oil demand and global oil supply. As with the various geopolitical risks, a peak oil scenario implies continually rising oil prices.

Projections of a peak in global oil production are highly controversial, for a number of reasons. The debate centres on four main problems: the lack of good quality oil reserve data, disputes over the methods used to calculate the likely timing of a global peak, ambiguity in the exact petroleum liquids to be included in the assessment, and disagreements over the ultimate impact of an oil peak. The first argument arises because the oil industry lacks a universally agreed set of oil reserves data, largely due to poor transparency and reliability of sources. Much of the oil reserve and field condition data is commercially confidential and thus well guarded by oil companies. At a higher level, many oil-producing countries view their reserve data as state secrets protected by their national oil companies. Public oil reserve data is sometimes fiddled or fudged for political or economic purposes, or manipulated to protect oil company profits.[19] In the late 1980s, for example, many OPEC (Organization of the Petroleum Exploring Countries) nations, in unison, reported dramatically increased oil reserves, a jump that was seen by many as a response to changed OPEC reserve-to-production quotas. In 2004 the Shell Oil Company sparked controversy and fines from UK and US financial regulators when it downgraded its reserves by 20 per cent. Even official reserves data collated by institutions such as the IEA or the BP Statistical Review of World Energy are viewed by many observers as error-riddled. This lack of clear and universally accepted data is in contrast to scientific consensus on other global challenges, such as climate change.

Second, and this is in part due to the lack of transparent reserves data, there is huge industry debate over the timing of a global oil peak, if it will happen at all. If global reserves are large, a global peak would occur much later than if global reserves are modest. And an earlier peak is more likely if reserves are much lower than is presently believed. Peak projections also rely on assessments of future discoveries and their timing. And there are methodological difficulties in assessing when peak oil might occur: no method has universal acceptance.

Third, there is also considerable debate as to which specific liquids or gaseous substances should be counted as 'oil'. For example, Canada holds massive reserves of sticky 'tar sands' which can be converted to crude oil, but at huge expense compared to the 'light sweet' crude which has traditionally formed the bulk of oil supplies.

While new 'non-conventional' sources might help boost overall oil supply, they will come at far greater cost than light sweet crude.

The result of these uncertainties is a flowering of analyses that have sought to predict the timing of a global oil peak. These range from November 2005[20] or 2007[21] to beyond 2030[22] or never.[23] One major review of 27 significant and credible oil production analyses, commissioned by the US Department of Energy,[24] found that 21 expected a global peak by 2020. It is unlikely that any consensus will be reached on the timing of peak oil, though, because the peak moment is also dependent on unpredictable variables such as global economic growth, oil consumption rates, petroleum production investment levels and geopolitical events. Some authors suggest that the actual occurrence of peak oil will only be recognised after the event, as global oil output sags despite rising prices.[25]

Finally, there is considerable debate over the potential impacts of peak oil. Some argue that a petroleum production peak will not be a problem because the world contains many undiscovered petroleum and other fossil fuel energy sources that better exploration and production techniques can exploit, especially as higher prices improve their economic viability.[26] Others suggest that production scarcity will dramatically increase prices and bring forward wondrous and innovative investment in production systems or stimulate new investment in efficient or alternative fuel technology that will moderate or minimise the effects of petroleum scarcity.

Some of the more pessimistic peak oil commentators suggest that petroleum is essential to economic growth and that declining supplies imply drastically reduced global economic capacity.[27] Such writers argue that neither the success of future exploration ventures nor the development of technologies that lead to energy independence from oil can be assumed. James Howard Kunstler's book *The Long Emergency* is perhaps the most pessimistic among this group, arguing that economies that are dependent on cheap and abundant petroleum will become dysfunctional and ultimately collapse.

Science, cult or consensus?

Uncertainty and the imaginative challenge posed by a world without oil have fuelled a huge literature on peak oil: there are countless books, newspaper and magazine articles, internet commentaries, weblogs and even documentary films. Many books have ominous titles, such as *Oil Crisis*, *The End of Oil*, *The Party's Over*, *The Last Oil Shock*, *Twilight in the Desert*, *The Long Emergency*, *Half Gone*, *Shortfall*, and *Beyond Oil*. Films about peak oil include *The End of Suburbia*, *Peak Oil: Imposed by Nature*, *Oil Crash: A Crude Awakening* and *Crude Impact*. Peak oil websites include www.peakoil.net, www.theoildrum.com, www.peak-oil-crisis.com, www.lifeafterthe-oilcrash.net or www.peakoil.org. Many offer doom-ridden warnings of social and economic collapse; others, such as the websites www.peakoil.net and www.theoildrum.com, present complex and detailed mathematical analyses of obscure and highly specialised oil, gas and coal data.

Despite denials of the concept of peak oil from many oil industry sources, there have been only a handful of books that are sceptical. Regrettably, these largely avoid detailed or rigorous critique of the theory. In *The Bottomless Well*, Peter Huber and Mark Mills argue that the world will never run out of energy because the pattern of history has been for human ingenuity to discover new energy sources and design more ingenious and efficient methods to exploit them. In *The Battle for Barrels*, Duncan Clarke eschews direct theoretical, methodological or empirical rebuttal of peak oil theory; instead he focuses on lambasting what he views as the misguided 'social ecology' of the peak oil 'movement'. While he throws some glancing blows at three peak oil theorists, he makes little positive attempt to bolster his claims of an unbounded world oil future.

One might have hoped that people in universities would have been thinking about such a critical strategic issue as oil depletion, but the academic sector has mostly dozed on peak oil questions. There are few formal academic research articles on the issue of petroleum constraints and peak oil. A smattering of papers and chapters in edited collections have emerged over the last decade, but these barely form a body of systematic inquiry. Kenneth Deffeyes'[28] *Hubbert's Peak* and *Beyond Oil* give a geology professor's view of peak oil, including a claim that the global production peak happened in 2005. Roger Bentley[29] provided a useful overview of the issue in an academic format while predicting a peak in oil production within five to ten years. D.L. Greene and colleagues concluded that:

> Peaking of conventional oil production is almost certain to occur soon enough to deserve immediate and serious attention. If peaking is already underway and oil supplies are as limited as the pessimists believe, the world is facing a drastic transition for which it is unprepared.[30]

Peter Odell, one of the few scholars who have considered peak oil in detail, is sceptical of the theory and describes the work of peak oil proponents as 'scaremongering'.[31] Nonetheless, Odell expects that conventional oil production will peak towards the end of the second decade of the 21st century. But his analysis doesn't take into account the increased voracity of China and India – his peak oil date could reasonably be moved earlier.

The lack of broad academic engagement with the question of whether and when peak oil will occur makes evaluation of the problem very difficult. In the absence of a clear scientific view, the task of officially evaluating the possibility of peak oil has fallen to governments and major oil industry players.

Official lines

If it were to occur, peak oil would pose such enormous problems for governments in oil-dependent countries that engaging publicly with the notion is fraught with risk for them. So most governments avoid assessments of future petroleum security and shy away from the mention of peak oil; they tend to be either sceptical of the theory or conservative in assessing its implications.

So any shifts in government positions on energy security and oil depletion serve as powerful indicators of the global trend in energy thinking, and the past two years have seen gradual but significant movement. Perhaps the most notable change comes from the IEA, whose task is to evaluate and advise on energy issues for its 27 member countries. The IEA held a very cautious view on energy security and oil depletion until recently. In 2004, it argued that:

> The world is not running out of oil just yet. Most estimates of proven oil reserves are high enough to meet the cumulative world demand we project over the next three decades. Our analysis suggests that global production of conventional oil will not peak before 2030 if the necessary investments are made.[32]

The IEA assessed the 'necessary investment' at US$3 trillion,[33] but this figure grew to $4.3 trillion[34] by 2006. By 2007 the IEA was rattled by continued high oil prices and rapid growth in China and India, and it then offered an implicit warning of the dangers of a production peak:

> Although new oil-production capacity additions from greenfield projects are expected to increase over the next five years, it is very uncertain whether they will be sufficient to compensate for the decline in output at existing fields and keep pace with the projected increase in demand. A supply-side crunch in the period to 2015, involving an abrupt escalation in oil prices, cannot be ruled out.[35]

This is a radical view for a traditionally conservative agency such as the IEA. The agency's chief economist, Fatih Birol, has since warned:

> Over the next few decades, our reserves of oil will start to run out and it is imperative that governments in both producing and consuming nations prepare now for that time ...[36]

Other government and official reports show increasing acceptance of declining energy security and the possibility of peak oil.

The Swedish government's[37] fear of peak oil by 2030 lies behind its plan to reduce transport oil consumption by 50 per cent by 2020. The UK Parliament has established an all-party committee to investigate peak oil issues. A 2005 US Department of Energy report[38] evaluated 12 significant and credible peak oil projections and the impacts peak oil might have on the US economy, then argued that planning to mitigate these impacts should begin immediately. The US military – one of the world's largest institutional consumers of petroleum – is also anxious about petroleum depletion. The US Army's Corps of Engineers has warned that:

> The doubling of oil prices from 2003–2005 is not an anomaly, but a picture of the future. Oil production is approaching its peak; low growth in availability can be expected for the next 5 to 10 years. As worldwide petroleum production peaks, geopolitics and market economics will cause even more significant price increases and security risks.[39]

Energy security and peak oil have also seen attention from the highest accounting authority in the United States. In early February 2007 the US Government Accountability Office (equivalent to the Australian National Audit Office) warned of the immense challenge that successive annual declines in global oil production would bring and of the need for rapid adaptation. The GAO called on the US government to immediately prepare a peak oil strategy.

Some major auto manufacturing and petroleum firms have also acknowledged concerns about petroleum security and peak oil. The Volvo motor company has given perhaps the most decisive corporate acknowledgment by admitting in early 2008 that 'global oil production will probably peak within a decade and the time of cheap and abundant crude oil will be over'.[40] General Motors' CEO Rick Wagoner has also been reported as admitting that peak oil poses a major challenge for the motor industry. The Chevron oil company became an early peak oil convert in 2005 when it said that 'the age of easy oil is over'.[41] The Chairman of the massive French oil company Total has confessed that 'the capacity of raising [oil] production is a real challenge ... if we stay with this type of production growth our impression is that peak production could be reached around 2020'.[42] Shell Oil CEO Jeroen van der Veer has warned that 'after 2015, easily accessible supplies of oil and gas probably will no longer keep up with demand'.[43] The world, he cautions, risks a dangerous scramble for control of oil resources.

Australian assessments of petroleum security

The tone of official Australian assessments has switched from complacent to anxious. A 2004 Australian energy White Paper set out a conservative yet cautious view of Australia's petroleum security:

> Despite increasing demand for oil, there are sufficient reserves
> to supply world demand for around 40 years ... In the longer
> term, concerns also exist about the longevity of oil supplies.[44]

Federal Energy Minister Martin Ferguson has since acknowledged that:

> With only about 8 years' known [domestic] oil reserves
> remaining at today's production rates, Australia is looking
> down the barrel of a $27 billion trade deficit in oil and
> condensate by 2015.[45]

An Australian Senate Committee investigated the future of Australia's oil supply and in early 2007 concluded that global oil production would likely peak by 2030. The committee warned that:

> [T]he possibility of a peak of conventional oil production
> before 2030 should be a matter of concern. Exactly when it
> occurs (which is very uncertain) is not the important point. In
> view of the enormous changes that will be needed to move to
> a less oil dependent future, Australia should be planning for it
> now.[46]

The Queensland government and the Brisbane City Council have also released assessments of energy security and oil vulnerability. The Queensland government report argued that demand would outstrip supply within 10 years and suggested 2013 plus or minus seven years as the possible timing for peak oil,[47] and is now preparing a peak oil mitigation plan. The Brisbane City Council didn't predict a date for peak oil, but the main projection cited by the report expects peak oil to occur by 2010.[48] The BCC report advocated for immediate planning for peak oil in conjunction with climate change mitigation and offered a set of recommendations, including a 50 per cent reduction in transport energy use. Few of these recommendations have been adopted as BCC policy.

Despite these attempts, Australian governments have not yet begun to recognise or address questions of energy security, peak oil and petroleum vulnerability in any systematic or deliberate way. There is no Australian government petroleum security strategy, nor do state governments have any plans to address energy security issues. The swag of state government urban plans brought in since 2000 largely ignore these issues. The only scheme to have acknowledged petroleum challenges is the Perth Network City strategy developed by the WA government.

End of an age

Australian cities are facing a future of energy uncertainty. We've presented a broad outline of the major current and likely future problems and challenges: demand crunches, geopolitical frictions and supply shocks. The risk of short-term shocks increases each day. The age of easy oil is well and truly over – we are racing headlong into an era of much higher fuel prices.

Such concerns are no longer held only by a small group of conspiracy theorists or dissidents. Governments and global corporations are gradually admitting that we face a troubled petroleum future. While we are sure that there will be many attempts to conjure magic new fuels and invent amazing new vehicles, as we point out in Chapter 6, one must have grave doubts about our ability to continue to deliver current levels of private mobility at current prices. If higher fuel prices are an inevitability, cities that have been built on the basis of cheap fuel will be transformed, whether they like it or not.

CHAPTER 4

Shocking the city: Higher fuel and housing costs in the suburbs

For those who've made their home on the edge of Australia's suburbs, life can seem tough. Hoppers Crossing is an outer tract of Melbourne's rapidly expanding western growth corridor. It's like many suburbs on the fringes of Australia's cities: people have flocked there to chase the promise of a better life, the sunny Australian dream of home ownership. For many, that dream has soured in recent years and the suburban sun has been shadowed by dark clouds. By mid 2006, escalating petrol prices and interest rates had forced at least one of Hoppers Crossing's residents to take a second job to keep his family's budget in the black. His neighbours have slashed their spending to make ends meet: holidays, entertainment, new appliances and health insurance have all been cut. 'It's just getting so hard,' he told the *Herald Sun* newspaper. 'They say we live in the lucky country but that's crap when people have to live like this.'[1]

Across Australia's suburban realm hundreds of thousands of households are facing the pressures of rising fuel, mortgage and grocery costs. Their cries of financial pain have sounded loud and clear across our cities. Relief looks out of sight. The spreading pain seems to jar with the idea of the lucky country and its bonny suburban landscape. If you believe the hype, economic luck seems all-pervasive today, with record low employment and the 17th year of an economic boom. Australia is riding on the back of a coal train to China. But as the 'dream' has always been steeped in fantasy, so the 'lucky country' was also a myth. Even as the resources boom reverberates across the nation, the latest wave of domestic dreams is shattering among the suburbs. Mortgagee possessions are spreading in their thousands within Australia's cities. How can petrol pain and mortgage stress afflict so many in a time of plenty?

To understand the fracturing of the dream and the likely trajectory of suburban Australia's economic and social prospects in the coming world of insecure global energy supplies we need to remind ourselves how the past few years of rising oil prices have tightened the screws on transport costs, living costs and debt.

Paying at the pump

Oil prices had stayed relatively low in the two decades up to 2004, in real US dollars, with the exception of a few sharp but short-lived blips due to events such as the first US war on Iraq. Since mid 2004, the price of oil has risen rapidly. While the oil price stayed between US$20 and $40 for most of the early 2000s, it began climbing after 2004. Since early 2007 the price of oil has rocketed continuously upwards. It climbed to $100 for the first time ever on 2 January 2008 and hit its highest ever level in early March. At the time of writing it was sitting at around $130 per barrel and still climbing almost every week.

Costlier oil has inevitably driven petrol prices higher in Australian cities. In January 2004 the price of petrol was around 92c/L across Australia. By January 2008 it was near $1.45/L:[2] a 50 per cent increase in 4 years. Whilst it is upsetting for motorists, the gain is surprisingly modest given that oil prices have more than doubled over the same period.

The 'bite at the bowser' has brought a clamour of complaint. Many families have found their fuel expenses draining increasingly large proportions of their budget. Transport costs make up around 16 per cent of household expenditure in Australia, and petrol makes up almost a quarter of all household transport costs, according to the last official comprehensive expenditure survey (in 2003).[3]

Table 4.1

Selected travel indicators for Sydney statistical division (SD) (2003)

		Travel Indicator					
		Average number of trips per person	Private vehicle mode share (all trips) (%)	Private vehicle mode share JTW (journey to work) (%)	Average trip length (km)	Daily VKT (vehicle kilometres travelled) per person (km)	Change in VKT per person (%) 1991–2001
Area	Inner/ East	3.85	48.7	49.2	5.7	10.1	-9.9
	North-East	4.01	67.9	65.2	8.2	17.9	0.3
	South-East	3.81	72.3	69.0	8.4	17.6	9.1
	Inner/ Central West	3.42	64.6	64.4	8.0	14.1	6.0
	North West	3.36	80.1	76.8	11.8	23.2	4.7
	South West	3.31	78.7	75.6	11.9	24.0	23.6
	Outer West	3.99	79.7	77.5	13.7	33.3	22.8
	Central Coast	4.16	77.3	77.3	12.8	30.1	19.0
	Total Syd. SD	3.74	70.0	67.6	9.5	20.0	11.6

SOURCE *Regional Transport Indicators for Sydney*, Department of Infrastructure, Planning and Natural Resources, NSW Government, 2005

The hardship of rising transport and fuel costs is spread unevenly; of course it hits those on modest or below-average incomes the hardest. However, the lowest-income fifth of households spend less than a quarter of the amount households in the top quintile spend on transport. It is households in the middle fifth which spend the greatest share of their incomes on fuel; they are followed by the next bracket down. These groups are the suburban 'working families' who traverse vast distances to work.

Perhaps the most significant feature of higher oil prices is how their impacts are spread across the Australian suburban landscape. As we showed in Chapter 2, the suburbanisation of Australian cities and the failure to extend public transport infrastructure has created vast tracts of urban car dependence. But not all areas are equally car dependent. Those most distant from public transport, and where homes and activities are dispersed, tend to be the most exposed. Sydney provides a particularly stark, but by no way unique, example (Table 4.1).

While residents of Sydney's outer western suburbs make around the same number of trips each day as those in the inner and middle areas, they are more tightly shackled to their cars. Around 80 per cent of outer suburban travel is by private motor vehicle compared with around two-thirds in middle areas and less than half in the inner eastern zones. Even worse, from an oil vulnerability standpoint, are the distances travelled by those in the outer suburbs; average outer suburban journey lengths are almost double those of middle and inner suburban residents, and average daily distances travelled by those in outer regions are nearly triple those of the denizens of the middle and inner zones. This urban transport gulf is growing: inner-city residents are shunning their cars and their trips are shrinking, while those in the outer suburbs wade further into the depths of car reliance. Similar patterns afflict other Australian cities.

Draining the tank

Even before the new stress of higher fuel prices, the cost of running a car was already a weighty financial burden. The 'whole of life' cost of running a car varies with the size and model of the vehicle. The average weekly running cost for a small car was estimated by the NRMA[4] in 2007 at $144 per week; for a medium car the average was $237; and for a large SUV the cost was $323. Petrol costs make up around 18 per cent of that cost, behind depreciation and the opportunity cost of vehicle capital value. An outer suburban family with a small car, a medium car and a medium 4WD could easily already be spending more than $600 per week in running costs for their fleet.

Consumers complain

So how have the residents of Australia's cities been responding to higher fuel prices? There has been little specific research into this, but a clear picture can be gained by piecing together evidence from many sources. Let us start with some simple consumer research. The Sensis market research agency has tracked public concern about national social and economic questions via quarterly consumer surveys since early 2004. After rising oil prices led the agency to add 'the price of petrol' to its list of consumer concerns in May 2005, households immediately rated this issue as their greatest fear, far above the conflict in Iraq, the environment and the workplace relations system. Petrol prices consistently topped consumer concerns during 2005 and 2006. While they dropped behind the drought and interest rates over a few quarters in 2006 and 2007, by February 2008 petrol prices were again the number one household fear. Terrorism, by comparison, barely rates as an issue of concern. Petrol is where the paranoia is.

According to the Sensis surveys, by September 2006, 45 per cent of households were having to slash their budgets to cope with higher fuel prices: 30 per cent were cutting back on entertainment and dining out, 22 per cent were trying to use their car less, 19 per cent were forgoing holidays and 11 per cent were cutting down on food. More than half of those on annual incomes of less than $35,000 reported cutting spending in an effort to cope. By February 2008 rising petrol prices were nominated as the greatest financial fear of households on annual incomes of up to $55,000. The Sensis surveys make it clear that households on the suburban edge are being shaken by petroleum paranoia.

The experience of those in Kellyville, on Sydney's northwestern edge, exemplifies the rising petrol stress. 'It is ludicrous,' one local family told the *Daily Telegraph* in May 2006, as their monthly household fuel bill rose by $80. 'Living in Kellyville we have no public transport, so [what we can do] is very limited. Something has to go. We can't just absorb it.'[5]

Close to busting

Another clue to the impact of higher fuel prices has been the dramatic growth in public transport use – for those who have access to it. Brisbane is a prime example of this massive boost. In each of the three years after 2004, when fuel prices shot up public transport use also jumped, by an average of 9.7 per cent, adding 12.3 million new passengers in 2005 followed by another 15.3 million in 2006.[6] This sudden shift took the region's transport planners by surprise – there had been just 3.8 per cent growth in the three years up to 2004. Bus services have struggled and ultimately failed to cope under the strain of such massive growth. Overcrowding has become widespread and chronic. In March 2007 a total of 1749 buses were forced to abandon passengers at stops because they were simply too full. Despite the addition of new buses by April 2008, 1800 services were still forced to leave passengers waiting at stops.[7]

In Melbourne, petrol pressures have hit the tram and train network. Melbourne's rail patronage has rocketed up by more than 10 per cent each year since 2004, from 137 million boardings to an estimated 200 million in 2008.[8] Overcrowding on the trains has been a source of intense public anger. The government and private operators have struggled to organise sufficient services to meet demand, and in March 2008 were forced to add another 200 services to the weekly timetable.

In Sydney, rail patronage had declined steadily during the first part of the decade, hitting a low of 270.3 million passengers by 2005.[9] After fuel prices jumped, rail patronage soared, with 11 million extra journeys boosting annual patronage to 281.3 million by 2007. This growth has been even more remarkable given the widespread dissatisfaction among travellers at the NSW government's management of the rail system.

Profound shocks have also been stressing Australia's automotive industry. Sales of large cars have slumped while light and small cars have been zipping out of showrooms. Large car sales in Australia plummeted from 203,524 in 2003 to 136,606 by 2006 – a drop of 33 per cent. Light and small car sales grew 51 and 25 per cent respectively. Australia's automotive industry faces a grim future. Poor sales of Mitsubishi's latest V6 car have forced the closure of its Adelaide plant, throwing a thousand workers into unemployment. Australia's remaining car manufacturers have for almost a decade been protected by federal subsidies and import tariffs. As consumers see the index of declining energy security turn over ever faster at the bowser they are downsizing their vehicle purchases. Australia's car manufacturers seem like dinosaurs, blind to the fiery comet heading their way as they continue to grind out big V6 cars.

Across the suburban plain the guzzling megafauna look suddenly unfit for the rapidly evolving and painfully selective ecology of petroleum insecurity. Cumbersome 4WDs have seen their second-hand values wiped out. At Chester Hill in Sydney's western suburbs the new fuel logic forced one of the local used car salespeople to slash the prices of 4WDs by more than 30 per cent to attract buyers. 'One guy traded his 4WD in the other day just because he couldn't afford the petrol any more,' the frustrated car dealer told the *Sydney Morning Herald*. 'Everybody seems to be looking for four-cylinders now because of the price of petrol.'[10]

The price of everything

While transport is where most households have felt the pointy end of the fuel spike, it is not the only place. Almost everything consumed by Australian households either contains oil or has been made or transported using oil. This means higher costs for other basic items, such as food and clothing, as the higher oil prices seep through the economy, creating a new era of inflationary pressure. Fuel inflation has obviously raced ahead since 2004 but oil pressures have pushed food inflation to more than double the general inflation rate. For the first time since records began, food prices are now decoupled from broader price movements.

The dependence of agricultural production on oil inputs and the new oil and food price link means this problem doesn't affect only Australia. Global food prices have rocketed since 2007, due to higher fuel costs and because of the sudden conversion of almost 5 per cent of the world's grain to biofuels such as ethanol and biodiesel. Food riots have broken out in many nations. Even the United States seems to be suffering from the new food shock – in late April 2008 reports of rice shortages, panic buying and hoarding appeared in US news media.[11] One commentator in the *Wall Street Journal* suggested that 'maybe it's time for Americans to start stockpiling food'.[12] Australia has been spared this kind of social disruption, yet reports abound of suburban families struggling to cope with the new food prices.

Hitting home

Ballooning inflation in Australia has set off changes in government policy. Under the Australian government's interest rate policy, the Reserve Bank of Australia is expected to keep inflation at 2 to 3 per cent by setting interest rates in a way that will control the supply of credit. Interest rates had slipped as low as 4.25 per cent in 2001 but the Reserve Bank has had to crank them up 12 times since 2002, to 7.25 per cent by early 2008. Oil has been a big factor in the eight of these interest rate hikes that have happened since oil prices took flight in early 2004. These increases have been transmitted directly to mortgage rates. Standard variable mortgage rates have climbed steadily from a 30-year low of 6.05 per cent in 2002 to 9 per cent by 2008. Most of this increase happened between 2005 and 2008, following a giant housing boom that inflated prices across the cities and propelled borrowing ever higher. And so we turn to the second great vulnerability that casts a shadow across Australia's suburbs: debt.

If the howling winds of rising fuel prices and higher living costs were already blowing a harsh winter across the Australian suburban steppe, the cold chill of debt is biting even harder into household dreams. Private household debt is the second great modern suburban vulnerability. The massive boom in home ownership in the 1950s and 1960s was heavily dependent on the availability of mortgage finance, and the past decade has seen vast escalation in household housing debt, whether for owner occupation or landlording. The new fringe estates and the remaking of older suburban landscapes are all built on borrowing. The wizardry of historically low interest rates has conjured the apparent miracle of affordable suburban home ownership. But it has also brought strange side effects that may yet turn into curses: contorted and hyper-inflated housing markets, over-geared and under-secured mortgages, super-sized houses and waves of outward urban spread.

Australia's housing debt in 1998 totalled $139 billion; by 2008 it had reached $448 billion. Such a binge was made possible by an unusual combination: low, stable interest rates and low unemployment. Mortgage rates had reached a high point in the early 1990s but steadily declined during the decade, hitting a low of 6.05 per cent by January 2002; this level had not been seen since March 1970. Thousands took the opportunity of the falling interest rates to race for the Australian dream, cheered on by banks, mortgage brokers, developers and politicians. This heady domestic reverie saw the proportion of Australian households with mortgages jump from 29.6 per cent in 1995 to 35.1 per cent by 2004.

The falling cost of debt also boosted borrowing power. Incomes increased as unemployment fell, giving households a new confidence about signing up for more debt. The ratio of housing debt to disposable income went from less than 50 per cent in the early 1990s to almost 140 per cent by late 2007. This is an average ratio, of course – for many across Australia's suburbs, housing debt far exceeds 140 per cent of their incomes.

This financial flood saw exuberant households flush with credit unleash a king tide in house prices across Australian cities. Median house values doubled from $250,000 to $500,000 in Sydney between 1998 and 2003.[13] House prices in Melbourne and Brisbane grew 91 per cent and 65 per cent respectively over the same period. First home buyers came spinning into the debt vortex. The average first home mortgage more than doubled, from $92,000 in 1995 to $209,600 in 2006. For those living in Sydney, the average first home loan was $260,000 by 2006. Not surprisingly, by 2003, 80 per cent of all Australian households paying more than half of their income on their mortgage were first home buyers.

The boom did not just inflate mortgages and prime house prices. The power formula of cheap debt and pumping dreams saw suburban housing bulk out into more and more squares of floor space. In New South Wales, the floor area of new homes spread by nearly a third between 1993 and 2003; it grew by a quarter in Victoria. Big cheap debts swelled houses into 'McMansions' set in utopian master-planned estates.

Few complained as interest rates collapsed in the mid 1990s. The few lonely wowsers at the property debt party were ignored. However, debts must be repaid. Money seemingly tamed may yet run wild. With interest rates gaining by more than a third since 2002, the dream is turning into a nightmare. The mortgage pain of the suburbs has finally stopped the great Australian barbecue. The revelry has ended.

By early 2004 house prices had stalled. Interest rates began to ratchet up and chew into household incomes. In March 1998, the average Australian household was spending 4.7 per cent of its disposable income on mortgage interest payments. A decade later this figure had more than doubled, to 9.5 per cent.[14] This is a greater proportion of income than at any time since measurement began in 1977. Some have had to restructure their debts: refinancing, as a proportion of new loans, rose from 19 per cent in 2000 to 31 per cent by 2008.[15] But people can only shuffle the deckchairs on their own personal *Titanics* for so long.

For some the pressure has been too great. Mortgage repossessions in Australian cities have begun tracking upwards. The banks are sending bailiffs in to take houses from defaulting owners. This is happening most in Sydney, where the tide of the housing boom lapped highest – and where its ebb is bringing the greatest pain. The NSW Supreme Court ordered 1750 writs of possession in 2004, at the peak of the house price boom; the figure in 2007 was 3935.[16] 'This is crazy,' one bailiff exclaimed. 'We don't need any statistics to know how bad mortgage stress is for people. We've got a whiteboard that we fill out and it's been chock-a-block for months.'[17] A 2007 study of mortgage delinquency by Fitch Ratings found repossession hotspots across western Sydney. Guildford was the worst affected, with 5.7 per cent of mortgages in default, followed by nearby Granville, Wetherill Park and Belmore.

Evictions are the scripts of suburban tragedy. One western Sydney real estate agent observed with dismay:

> It's very, very sad. We've had some cases where they've handed
> the keys over and they're still wiping the kitchen down. They're
> still proud of their home.[18]

Some have been cut deeply by the socio-economic shards of splintering dreams. Another western Sydney realtor described one case:

> The wife was under the impression that it had all been sorted
> out the day before. But then the sheriff arrived with the
> repossession order. She didn't really understand what was
> going on and was quite hysterical. She had four children. It was
> tough.[19]

Shocking the suburbs

Petrol prices appear set to continue their upward climb, and inflation pressures are increasing household bills. Even if mortgage rates
settle at their current levels, strong inflation and rising oil prices
will perpetuate the ache of debt. Over the longer term, oil prices are
likely to overtake mortgage debt as the greatest household stressor.
But not all suburbs are equally affected. And if we are to craft policy
and planning responses to these needs we must first ask which areas
will be hardest hit.

Over the past three years we have been investigating petrol and
mortgage vulnerability and charting the problem across Australia's
urban landscapes. To do this we created the 'Vulnerability Assessment for Mortgage, Petrol and Inflation Risks and Expenditure'
– the VAMPIRE. This is an index which calculates the level of
household vulnerability. It is based on Australian Census data. It
combines information on car dependence, mortgages and incomes
at the 'collection district' level (about 200 households). Space
prevents greater methodological discussion here but those with a
statistical bent may peruse our earlier paper for further descrip-
tion.[20] When the 2006 Census data recently became available we
updated our 2001 VAMPIRE analysis. The broad patterns of oil
and mortgage vulnerability are described in the following sections.
Unfortunately we can't include the VAMPIRE maps of Brisbane,
Sydney, Melbourne, Adelaide and Perth here; those wishing to view
them should consult our other work.[21]

Across all these cities, several key patterns are apparent – each
is described below.

Minimal and Low Vulnerability Areas

Inner-city areas of these five Australian cities almost universally
fall into low or moderate vulnerability categories. Places like New
Farm and Indooroopilly in Brisbane, Crows Nest and Surry Hills in
Sydney, Fitzroy and St Kilda in Melbourne, inner-north and inner-
eastern Adelaide and Crawley, Claremont and central Perth all rate

as having low oil and mortgage vulnerability. Residents of these areas are wealthy, use public transport often, and walk or cycle more than those further from the city centres, in part because these areas have some of the best public transport services. Some areas distant from the CBD also have minimal or low oil vulnerability – Chapel Hill and Kenmore in Brisbane and most of Sydney's northern suburbs (east of Epping). Low oil vulnerability areas are also found as far east as Box Hill in Melbourne and as far southeast as Blackwood in Adelaide. Some areas along Perth's western shorefront, such as Mulalloo and Scarborough, also fit into this category.

Moderate Vulnerability Areas

The middle suburban rings of Australian cities are more oil vulnerable than those at the centre, but only moderately so: Hamilton, Stafford and Holland Park in Brisbane, Sydney's Parramatta Road corridor and middle southern suburbs, such as Canterbury, plus almost all of northern Sydney between West Pennant Hills and Chatswood. Melbourne's moderate oil and mortgage vulnerability middle suburbs extend from Essendon in the west through Brunswick and Northcote to Doncaster then south to Brighton via Surry Hills and Chadstone. In Adelaide a band of moderate oil vulnerability localities west of the central city extends south from Marion to Mitchell Park and immediately northwest of the Adelaide CBD. In Perth moderate vulnerability areas are found to the south, east and north of the CBD in places such as Bull Creek, Cloverdale and Duncraig.

High Vulnerability Areas

In Australian cities large areas of moderate to high oil and mortgage vulnerability are found in the middle and outer suburbs. In Brisbane these areas include suburbs such as Narangba, Beenleigh, Karalee and Drewvale, all in outer suburban areas. Sydney's areas of moderate to high oil and mortgage vulnerability include outer suburbs such as Campbelltown, Regents Park, Penrith and

Camden. In Melbourne there is a large swathe of high oil vulnerability suburbs, including Sunshine, Lilydale, Werribee and South Morang. Adelaide's high oil vulnerability zones include suburbs such as Northfield, Paralowie, Craigmore and Campbelltown. In Perth, high vulnerability areas are found among outer suburban zones such as Coogee in the south, Bassendean and Kenwick in the east, plus Innaloo, and Karalloo in the north.

Very High Vulnerability Areas

Very high vulnerability affects large tracts of the outer zones of Australia's cities. These include places such as Caboolture, Redcliffe, Capalaba and Browns Plains in Brisbane. In Sydney high oil and mortgage vulnerability is distributed across much of the city's western suburbs, including Hebersham, Green Valley and Cabramatta in the mid and outer west. Many parts of Sydney's middle west, such as around Bankstown, also show high levels of oil vulnerability. In Melbourne the pattern of outer suburban oil vulnerability is particularly pronounced, with many high vulnerability zones in outer suburban growth corridors such as Frankston, Cranbourne, Berwick and Knox Park in the southeast plus Epping, Roxburgh Park and Craigieburn in the north and Kings Park, Deer Park and Hoppers Crossing in the west. In Adelaide the most vulnerable suburbs include Queenstown, Burton, Holden Hill and Elizabeth in the north, plus Christies Beach, Hackham West, Morphett Vale and Sturt in the south. Perth's most vulnerable localities include Banksia Grove, Girrawheen, Balga, Marangaroo, Alexander Heights, Ballajura, Beechboro and Morely in the north, Gooseberry Hill in the east and Langford, Huntingdale, Seville Grove, Armadale, Hamilton Hill and Success in the south. Oil and mortgage vulnerability is overwhelmingly an outer suburban challenge.

On the edge

The conclusion to be drawn from the VAMPIRE is that urban structure heavily determines the extent to which households are likely to feel petrol and mortgage stress. It shows that across Brisbane, Sydney, Melbourne, Adelaide and Perth it is the outer suburbs that are likely to suffer most in the coming energy and credit barrage. The middle and inner zones are far less vulnerable. This is of course no surprise, given the history of Australia's cities and suburbs and the spread of income and mortgage pressures.

Despite the continuing popular belief that Australia is an egalitarian society, the VAMPIRE shows, as have countless other studies, that our cities are scored by severe socio-economic divisions. Since the 1960s the outer suburbs have been the places where those on the lowest incomes have been directed by housing and labour markets; the wealthy have retaken the historic city cores, with their abundance of jobs, services and public transport links. Residents of centre and middle suburbs will have little difficulty switching to public transport, and their higher incomes will shelter them against the mortgage winter. Those in the outer suburbs will have fewer viable choices over their travel, though the distances they must traverse are typically much greater. Many are also perched on the outer rim of the housing market, where house prices are at greatest risk of receding. A mortgage winter could leave many frozen on the edge.

In a world where the cost of petroleum is rising almost by the week, the bleak divisions of the VAMPIRE appear set to produce a deeply unfair set of results. Those in the high vulnerability VAMPIRE areas are among the least wealthy, but will face the biggest increases in living costs. Most have highly constrained 'choices' in reaching for the Australian dream: their income limits their housing purchase options to the outer suburbs, and car dependence is the unpleasant corollary.

The wide expanses of car-dependent suburbia that encircle Australian cities are no accident. Accumulated government decisions since the 1950s are the ultimate cause. The problems are well known and the VAMPIRE effect represents only their latest manifestation. Yet governments have shown a profound reluctance to alter their planning to ensure suburban resilience in the face of oil and mortgage stress. This may change, though, because when the residents of Australia's suburbs feel the VAMPIRE's bite they are unlikely to suffer in silence. Cracks have appeared in the suburban veneer. Political tradesman will soon be called to fix them. The next chapter examines the politics of petrol and mortgage vulnerability.

CHAPTER 5

Oil and politics:
The impact of rising fuel costs
on suburban constituencies

As the price of oil began its ascent in the mid 2004, Australia's political climate seemed settled, if uncomfortable for some. John Howard, as Prime Minister and Liberal Party leader, had been in power for eight years. After some faltering steps early in his tenure he was seen as near invincible. Howard's rise to power was in part due to the support he won from disenfranchised post-industrial 'battlers' in the middle and outer suburbs. This loosely defined class was made up of the skilled trades and semi-skilled service workers who had struggled out of the recession of the early 1990s to take advantage of the new suburban housing and service economy. Howard counted among his electoral strengths the ability to attract a significant cohort of these households. After political commentators and the media 'discovered' this group, the group became known as 'aspirational' voters.

At the individual electorate level aspirational voters have displayed a profound ambivalence to both the social concern of Labor and the privatism of the Liberals – this has caused no shortage of discomfort for both parties. It translated into an electoral landscape in which small local swings could exert great influence on national outcomes. In the absence of large swings in other seats, small shifts in outer suburban marginal seats have influenced elections over the last decade. Candidates from both major urban parties have vied for advantage in these outer suburban marginal seats.

The rhythm of invincibility Howard had achieved by the mid 2000s allowed many observers to forget the astonishing political turnaround he had engineered in the early years of the decade. In March 2001, leaked internal party notes revealed that the Howard government was seen as out of touch, 'mean and tricky', and was facing political rejection. The turning point in Howard's political fortune came in the by-election held for the Melbourne seat of Aston in August 2001. Aston sits in the outer-eastern suburbs of Melbourne, where public transport is poor and lower middle-income households were heavily indebted from the home ownership boom. The by-election was triggered by the death of a locally popular Liberal, and the seat was viewed as nominally 'marginal'.

Despite Howard's woes, Labor ran a poor campaign in Aston under the leadership of Kim Beazley. Labor's candidate was a junior senator's adviser who had been parachuted in from across town, and he wilted in the glare of the national spotlight. Labor was also distracted and unsettled on environmental issues by the Greens and by a ragtag miscellany of single-issue campaigners, including a clutch of transport activists running under the banner 'Public Transport First'. The Liberals, in contrast, nominated a sober local accountant and family man who conveyed a sense of connection with local concerns. The Liberal Party won the by-election and Howard was returned shortly after in the November 2001 general election, which saw Labor relegated to its third hard stint in opposition since 1996. The new Aston member clearly recognised the foundations of his success in the newly created matrix of suburban home ownership and debt:

> One of this government's greatest achievements has been the
> management of our economy and the subsequent reduction
> of mortgage interest rates. Mortgage interest rates peaked at
> 17 per cent under Labor. Today they are around 6 per cent.
> This means that home ownership – the great Australian dream

– has become a reality for many more thousands of Australians.
It seems to me that we often forget about these enormous
achievements that help so many across the community, and this
is precisely what this government is all about.[1]

The Liberals clearly learned from Aston. The home ownership and
interest rate theme was taken up with gusto by Howard in the next
election, which pitted him against the newly installed Labor leader
Mark Latham. The 2004 election came as interest rates had begun
their ascent and oil prices, while still relatively low, had begun their
upward journey. This election was significant from the standpoint
of suburban oil and mortgage vulnerability, because it prefigured
the battleground of the 2007 poll, which Labor won.

In Latham, Labor had chosen a leader who told stories of aspi-
rationalism and who claimed to understand this new class of suburban debtholder. Latham used his own electorate of Werriwa as an
example of the new political genus:

When I grew up … our values were based on the politics of
us versus them – the working class versus the old money.
Now when a person grows up in my electorate, they can see
prosperity in the neighbourhood next door. Social mobility has
become more tangible and achievable. The politics of envy has
been replaced by the politics of aspiration.[2]

Despite Latham's affinity for this newly discovered fraction of the
suburban populace, Labor lost the 2004 election. This was in part
due to Latham's profound misunderstanding of the basis for the
aspirational suburbanism that he trumpeted. He believed in and
articulated the notion of a confident, self-made outer suburban
middle class ascending a 'ladder of opportunity'. In doing so he
presumed a high degree of agency among this group, and down-
played the shaky foundations on which their apparent success was
built. He unintentionally contradicted his image of a self-assured
suburbia with his repeated promise to 'ease the squeeze' of mount-
ing financial stressors. While Latham courted and flattered the

suburbs, he could not craft a message that could both celebrate confidence and warn of insecurity ahead.

Howard, in contrast to Latham, seemed to recognise that the McMansion symbolised the underlying fragility of the suburban debt matrix. Indeed the consumption of services which was inherent in the aspirational outlook was heavily underwritten by Howard's 'secret Keynesianism'.[3] His government fed hundreds of millions of dollars of taxpayers' money to households for the consumption of private services, such as health insurance and schools. Residents of the outer suburbs, it seems, proved a more complex and vulnerable group, for whom newfound social status was accompanied by deep anxiety over the fragile structure of their indebtedness.

In Aston, Howard had paraded the gifts of low interest rates and home ownership that he had bestowed on the suburbs. But by 2004 he warned of their confiscation by Labor. Howard raised the interest alarm across the fringe marginal seats and claimed that with Latham prowling their neighbourhood, suburban households risked the loss of their valuables. With Howard badged as chief neighbourhood watcher, Latham seemed more like a suburban bogeyman than a comforting bedtime storyteller.

Losing the suburbs

Howard's 2004 win relied heavily on his interest rate promises, so he was snap frozen by the cold inflationary winds that swept across the electoral landscape in the following three years. Suburban tension over debt and the price of petrol was a throbbing pain for which he had no salve. The dramatic increases in oil prices between 2004 and 2007 pushed him far into political arrears. The global oil price sat at US$51 per barrel the day before the 2004 election but had nearly doubled, to $95 per barrel, by election eve 2007. Petrol prices over the same period increased by almost a third, from AUD$1.04/L to $1.36/L.[4] As petrol prices started pushing inflation up, the Reserve Bank began pushing interest rates up. The mortgage screws started turning for suburban households.

The Howard government was slow in sympathising with burdened suburban households. Treasurer Peter Costello mused about a 'third oil shock'. Howard warned increasingly worried motorists that petrol was unlikely to fall below $1 per litre again.[5] By mid 2006 the pressure was unbearable: in the Budget Howard and Costello claimed to have spun a tranche of tax cuts for the wealthy into crucial help for modest income households:

> People are paying a lot more for petrol and one of the best ways
> to help people with the high petrol prices is to put money in
> their pockets … The answer is to put more money into people's
> pockets because the high price of petrol is having a depressing
> effect on people's livelihoods, people's incomes.[6]

If Howard was serious about relieving the petrol burden on households, tax cuts were a weak and indirect means of doing it. The new tax scales gave greatest benefit to wealthy households rather than those on modest incomes. The political alarms were sounding – Howard and the Liberals were under pressure from higher fuel prices and their effect on mortgage interest rates. Howard's attempt to recast the labour market with his WorkChoices Act only brought further anxiety. Suburban voters were getting nervous.

Howard was as impotent in quelling the rising suburban clamour as he was in countering the rise of Kevin Rudd to the Labor Party leadership. Labor had learned from 2004, and with costly petrol and biting debts, Howard's aspirationals were once more transfigured by Labor's rhetoric into 'working families' doing it tough. As Rudd argued:

> Despite the big picture on the economy appearing good, under
> the surface many family budgets are simply stretching to
> breaking point. Mr Howard says working families have never
> been better off, but housing is becoming less affordable and
> taking more and more of the family budget. Mr Howard has
> broken his promise on interest rates – we've now had nine
> increases in a row – and the rising cost of petrol, groceries,

childcare and the loss of penalty rates and overtime under
WorkChoices [means] people are asking, 'If the economy is
doing so well, why aren't we?'[7]

The Liberals tried various methods to subdue the suburban hue
and cry, including an inquiry into petrol prices – one of many over
the past decade. But such symbolism failed. With every interest rate
rise Howard was being hoisted on the political petard of his 2004
victory. Yet, curiously, the national statistics all looked dazzlingly
good. Australia's GDP growth remained solid; unemployment was
at 30-year lows. China was buying anything that could be dug and
burnt.

But the mining boom was also having perverse effects. Large
parts of Australia's economy sped faster on the hurtling outback
coal trains, but those in the suburbs felt left at the loading dock.
The same economic transformations that opened the door to the
minerals boom were convulsing global energy markets and fuel
costs. Howard seemed dazzled by the lights of the coal trains and
the armada of bulk carriers and was unable to see the rising levels
of fractious suburban anxiety.

It is not surprising that neither the Coalition government nor
Labor foresaw the post 2004 global energy price rise. Most official
forecasts in the mid 2000s expected prices to fall, not rise. In December
2004 the government's official forecaster ABARE (the Australian
Bureau of Agricultural and Resource Economics) had predicted
oil prices slipping to US$35 per barrel in 2005, but instead they
increased to $57 per barrel.[8] ABARE also forecast that 2007 would
bring $56 per barrel crude – oil prices jumped defiantly to $81 per
barrel. With the oil shock so woefully underestimated by economic
soothsayers, it is no surprise that the government didn't respond
more emphatically.

Howard's promises proved unpersuasive. He also battled restive
labour and untamed nature. Unions stirred suburban workers
against a chilling industrial regime, and public awe of drought and
global warming brought a new heat to Australian politics. In the

November 2007 election, the Liberal-National coalition lost office, shedding 23 seats to Labor, including a spectacular repudiation of the Prime Minister in his own seat of Bennelong. The battlers set aside a decade of aspiration and relearnt the hard work of suburbia.

The VAMPIRE and the voters

Labor was carried to office in 2007 on the shoulders of its newfound suburban 'working families'. Many Liberal seats across the outer suburbs were lost. In places like Longman, Forde and Blair, on Brisbane's outer fringe, crowds of voters thronged to Labor. In Sydney's west, the electorates of Lindsay, Parramatta and Bennelong all fell Labor's way. The oil and mortgage vulnerability dynamic was palpable. Many of these zones were where the VAMPIRE was biting hardest.

We've looked at this effect by calculating the two-party preferred swing in voter sentiment booth by booth relative to our maps of oil and mortgage vulnerability. This analysis shows that oil and mortgage vulnerability is scouring new contours in the electoral sands.[9] Sydney is a good example.

The results for Sydney show that most polling booths across the city swung towards Labor. Only a very small proportion shifted towards the Liberal Party. The size of the swing seems to be associated with underlying VAMPIRE patterns. In general, the booths that saw very slight swings to Labor, or swings to the Liberals, were in Sydney's eastern suburbs – such as in the electorate of Wentworth. These booths were mostly in areas rated as having low oil vulnerability on the VAMPIRE index. Booths in much higher oil vulnerability zones, such as Sydney's middle west, southwest and northwest, largely saw big swings to Labor – many shifting by at least 10 per cent on a two-party preferred basis. These results suggest that the level of oil vulnerability is altering electoral behaviour, with higher vulnerability zones tending to swing more heavily to Labor than less vulnerable areas. Many suburban areas which had for a decade supported John Howard's Liberals jumped Labor's way. There were some exceptions, though: a scattering of booths in the highly oil and mortgage vulnerable areas around Liverpool and Fairfield swung very slightly to the Liberals, perhaps because they were already among Labor's firmest supporters. The broad electoral effect of increased oil and mortgage vulnerability in Sydney has been to push voters towards Labor. Similar effects are apparent in Brisbane and Melbourne.

A new suburban political landscape

As the experience of John Howard demonstrates, politicians who fail to navigate the new suburban landscape of oil and debt will find themselves badly off course or, even worse, lost in the wilderness. If petrol hit $2.00 per litre or higher, Australia's car-dependent and mortgage-burdened aspirational voters would quickly become desperational voters. Any party seeking office in the future will need to deal with this new energy security reality. The petrol policies announced by the parties so far have been mostly invective and symbol. There are few signs that either major party is treating petroleum with the seriousness it deserves.

Talk of relief to offset the oil, food and mortgage stresses has, however, already entered the political conversation for both major parties. Both have so far fluffed their lines. Labor is set to introduce a 'FuelWatch' scheme which will require petrol vendors to register a set price each day rather than adjust at will. FuelWatch will have little, if any, discernible effect on petrol prices at the national scale – galloping global oil prices will easily swallow the 1.9 cents per litre the scheme is hoped to shave from the bowser price. Even the Australian Competition and Consumer Commission (ACCC) has cast doubt on the ability of FuelWatch to suppress petrol prices.[10] Labor must do much better than FuelWatch.

The Liberal Party proposal to cut fuel excise by 5 cents per litre is at least as deficient. Cutting excise by this amount would strip $1.75 billion from government revenue,[11] but would provide minuscule and poorly targeted relief. Indeed, the wealthiest households, who tend to spend more in absolute terms on petrol, would receive the greatest benefit from a fuel excise cut. Such direct suppression of petrol costs will only mean a crueller adjustment later.

Political attempts to fix petrol prices through regulation or tax fiddling are doomed to failure because the government controls almost none of the factors that are driving international energy prices. Our political leaders will only make their problems worse by telling voters that they can alleviate the sting of petrol prices by playing big brother at the bowser and tweaking taxes. The rising cost of energy is a sharp signal that the current suburban matrix of debt-dependent housing and oil-reliant transport may not endure. Our political leaders need to start rethinking not just the functional basis for Australian suburbia, but its political basis as well. Any political program that seriously tries to deal with the politics of energy constraint must seek to reconstruct our oil-vulnerable suburbs so that they can be sustained in a period of continued petroleum pressure. The political challenge is to redistribute transport choice in a time of scarcity.

The lost freeway of choice

A household's ability to choose viable alternative travel modes will soon shape its social and economic status. Suburban households, in the main, lack the choices in transport that can allow them to leave their cars at home. Our politicians need to expand the viable transport mode choices already found in the inner cities to the middle and outer suburbs.

The post-World War II period saw rapid suburban expansion based on the automobile, and government transport spending fell heavily. Governments saved large sums of money by deciding not to run public transport to the new suburban housing tracts. Instead, they ploughed funds into freeways. The political program underpinning Australia's postwar suburbia wrapped new notions of personal freedom, unbounded mobility and travel choice into a 'love affair' with the car. But the romantic liberal freedom of suburban travel choice has been illusory for many. Suburbs built around the car have always offered poor travel choices for the poor, the aged, the young and those unable to drive. Many suburban households are effectively forced by lack of transport alternatives into car owner-ship.[12] Higher petrol prices make the car more cruel domineer than romantic liberator. The love affair is facing trouble and strife.

The idea of travel choice is popular with those who actually have it – those in the parts of Australia's cities which are arguably more 'liberal' (small-L and large) than the suburban hinterlands. Wealthy inner urban suburbs set among Liberal Party bastions, like Waverley in the seat of Wentworth or Lane Cove in North Sydney, are among the greatest public transport users in Sydney: 42 per cent and 43 per cent (respectively) of travel in these suburbs is by public transport or walking.[13] Wentworth MP Malcolm Turnbull has been an enthusiastic public transport traveller and even lobbied for permission to spend his parliamentary allowance on a yearly public transport ticket because, extraordinarily, he finds it 'more efficient' than a car for travel around his electorate.[14] But millions in the suburbs do not have this option.

Our car-dependent suburbs are about to crash headlong into the new politics of transport choice. Those politics are as yet indistinct and poorly formed and their expression has come mainly through complaints in media grabs or talkback radio. Public appreciation of the global energy crunch is still developing and doubts over petro-leum supply and security have not yet entered broad community consciousness. Before long, however, realisation will dawn about the implication of our suburbs' car dependence and lack of trans-port choices. Blame will be attributed and solutions sought. Our politicians and policy makers face the task of improving the quality of public transport in the middle and central zones of our cities and out to the suburban areas.

Planning beyond the freeway to serfdom

Planning has long been out of political vogue in Australia. Metropolitan planning has withered among the states. The federal government has not had a serious urban policy with a suburban focus since the era of the Department of Urban and Regional Development in the 1970s. The Better Cities Program of the early 1990s emphasised inner urban redevelopment at the expense of planning at the metropolitan scale. Since 2002, however, almost all state governments have introduced metropolitan strategies, beginning with the contentious *Melbourne 2030 Metropolitan Strategy*. The quality of these schemes has been highly variable; their reception, and their implementation, have equally been mixed. Most have suffered from the diffidence of their sponsoring governments. The failures of the *Melbourne 2030 Metropolitan Strategy* have been heavily debated in that city, with particularly trenchant criticism reserved for the inadequacies of the plan's transport components (or lack of them), including in relation to the issue of suburban oil vulnerability.[15]

There is one area of metropolitan policy, however, where state planning has reigned supreme for decades. Our cities' road planners have been steadily and meticulously rolling out ever bigger and wider freeways and, latterly, tollways. Slowly but surely, the motorway schemes drawn up in the 1960s have been connected, assisted by lavish state and federal assistance. The extreme car dependence – and supposed freedom of the automobile – of Australian cities has been no accident. While we are now desperately uneasy about the future of petroleum supplies and the price of fuel, the triumph of the road planners at least shows us, ironically, that a long-term state-driven program of public investment can be wildly successful. We need to apply this same meticulous and determined mindset to redistributing public transport across the Australian metropolis.

Proposals for massive new state-led investment in public transport will no doubt be scorned by the priests of economic rationalism, who have held political sway in Australia for the past two decades. They will insist that the market god will reshape our suburbs for the best of all possible oil-constrained futures without need for government-led planning. They are welcome to their views and the impotent passivity they imply. The suburban congregations stuck on the freeway to serfdom are likely to be far less sanguine about their transport future. With fuel prices headed up into the unknown, no politician with an eye to electoral longevity will be able to avoid the question of government planning to reduce the oil vulnerability of our suburbs. At the time of writing, this lesson was being learned the hard way by Labor and Liberal alike as they clashed over the regulatory trivia of FuelWatch and petrol excise cuts; parliamentary pantomime is no substitute for comprehensive planning.

Public transport is among the most intensely local of public services but it must be planned in detail if it is to work effectively. The vague promises to improve suburban public transport found in most contemporary metropolitan transport plans will not suffice; nor will affirmations of the beneficence of private transport providers. As fuel prices march ever higher, suburban households will be demanding new high-quality public transport within a short walk of their front door. They will demand that their street and suburb be linked into the regional public transport system. They will make these claims plain to their local MPs and the electoral results at the booth level will allow us to track the results at the next election. Both major parties have time to build up their oil vulnerability credentials before then, but the race is on – who will win the battle of the bowser and the buses that will define Australia's urban electoral future?

A way without will?

Labor is perhaps best positioned politically for the rediscovery of centralised planning and funding of public transport, but its disdain for such policies is barely distinguishable from that of the opposition. In some states Labor and Liberal policies on transport are almost identical. However, at the time of writing all states are Labor controlled and most of the metropolitan plans produced in recent years have been Labor documents. Labor also has many stains on its record in transport and metropolitan planning. Reckless urban freeway expansion over the past decade, especially in Melbourne and Sydney, has only entrenched suburban car dependence; neglect of public transport has been deep and deliberate.

Yet among Labor's ranks are some of the politicians who have shown greatest leadership in reducing oil and car dependence. WA Planning and Infrastructure Minister Alannah McTiernan has a long record of advocacy on energy security and sustainable transport planning issues, and oversaw construction of a 70km rail line through Perth's southern suburbs. Queensland Sustainability Minister Andrew McNamara is known internationally as a strong advocate on energy and peak oil issues; one of his prescient early achievements was to persuade former Queensland Premier Peter Beattie to establish Australia's first Oil Vulnerability Taskforce. Unfortunately for Australia's cities, McTiernan and McNamara are rare voices. They and their less vocal like-minded colleagues need to be given a greater voice in Labor debates on energy security and urban transport.

The antipathy for public planning shown by the Liberal Party has positioned it poorly for the coming period. In Victoria, the Kennett Liberal government inflicted terrible damage on public transport through privatisation (which Labor quixotically supports despite clear evidence of its failure).[16] Few prominent Liberal Party office holders have made any attempt to seriously address either suburban car dependence or declining energy security. Federal Shadow Treasurer Malcolm Turnbull has come closest through his advocacy on public transport, but his efforts barely indicate an energy security perspective, let alone a significant vision for reconstruction. The Liberals will have to either begin to support state-led public transport planning or provide a detailed explanation as to why we should not do it.

If Australian cities and suburbs are to survive rising fuel prices our governments must reject the model of urbanisation pursued for the past six decades. They must start planning to ensure that higher fuel prices do not stunt our cities' growth by ruining the suburbs. Does either of our major parties have the leadership necessary to face the energy security challenge, to explain to voters the difficult reality about the world's energy supply and to actively and competently plan for the protection of our suburbs? Finding this leadership will be the difficult part of this process. As we show in the next chapter, Australia has an abundance of ideas for the content of oil vulnerability plans.

CHAPTER 6

Oil-proofing Australian cities: Transport and planning solutions to oil vulnerability

The strategic thinking that has been done in Australian energy and transport policy has focused on technological solutions: new fuels that will be able to replace petroleum, or new vehicles that will be so energy efficient that they will allow us to sustain today's level of unbounded mobility, will be discovered or developed. We are sceptical of this reliance on new technology. There is little evidence that any emergent new fuel will match conventional light sweet crude oil on energy density, portability, ease of production and, most important of all, price. Nor is there much evidence of any vehicle arriving in the near future that will both run on a new magic fuel and be affordable.

Many economists, including those who advise our governments, believe that high market prices for petroleum will create such powerful incentives that private firms will scramble to develop new fuels and technologies. The former Director of ABARE famously told the 2006 Senate Inquiry into Australia's energy security: 'If the price of eggs is high enough, even the roosters will start to lay.'[1] Unfortunately, the evidence on which such reasoning is based is conjectural, and the analogy is absurd. Adapting this analogy to petroleum, the problem is not so much the achievement of major fuel and vehicular technological advances *per se*, however doubtful that is, but the cost at which any such technology might become feasible for our car-dependent suburban households. The global oil price needed to get the petroleum roosters a-laying is likely to be far in excess of the fuel prices on which Australian suburbia was founded.

Also, leaving it to the market to figure out our suburban transport future is no solution. To allow such an enormous if to hang over our suburbs is hardly good policy. Even assuming there were evidence that new technologies that might provide the same private mobility as the car could be developed, it is unlikely that these could operate at the same mass scale as our current fuels and vehicles and, most importantly, at the same price. Even the ABARE Director's optimism was undermined by his own – commendably honest – revelation that 'there's no doubt that I have made the occasional mistake with my oil price forecast and quite a few other forecasts, frankly'.[2]

Addressing the petroleum security challenge means taking an 'evidence-based' approach to reconstructing our suburbs. The first half of the 20th century showed us that in the absence of mass auto-mobility and cheap abundant petroleum, our cities still grew, and became spectacular metropolises. Australian cities achieved such feats almost totally with public transport, as did many cities across the globe, particularly in Europe. Before we explain how this works, it's worth reviewing why some of the vaunted alternative fuels and vehicles will fail to rescue our suburbs.

Looking for oilternatives

We don't have the space for an exhaustive review of all the weird and wonderful technologies that purport to solve the urban travel chal-lenge; nor are we fuel or vehicle technology experts. However, we can present some of the major possibilities and difficulties with the main candidates and assess whether they can provide the current levels of mobility at the very low fuel prices on which our suburbs were founded.

The first set of options draws on alternative fossil fuels as substitutes for conventional crude oil – natural gas, coal, tar sands and oil shale have all been nominated.

Gas? Naturally

Natural gas burns cleanly, releases modest greenhouse emissions and can be relatively easily converted for use in motor vehicles. To convert Australia's 14.4 million motor vehicles at a cost of around $2000 per vehicle would cost $28.8 billion. Even if this were done, mass natural gas use in motor vehicles is problematic. Natural gas is volatile and the distribution infrastructure required for its widespread use in motor vehicles doesn't exist. The economics are also challenging. A large and sustained rush to natural gas for transport would throw motorists into competition with existing agricultural, residential and commercial gas users. This competition would also be global, given that other economies will also be searching for petroleum alternatives. Natural gas also suffers from a depletion profile similar to that of crude oil, with some authors suggesting that 'peak gas' could quickly follow peak oil.[3] The technical problems of mass natural gas conversion of our vehicle fleets are probably insurmountable within the timeframes predicted for the demand pressures on petroleum supplies. Even if these obstacles could be overcome, mass use of natural gas would still see us paying dramatically more for urban car use than we currently pay.

King Coal to the rescue?

Because coal can be converted to a gas or liquid fuel via the 'Fischer-Tropsch' process, which Germany used in World War II, Australia's coal resources could be used to produce transport fuels. The federal government has already begun to explore 'coal-to-liquid' possibilities. Unfortunately, coal liquification uses much more energy than petroleum production does. This means higher greenhouse gas emissions. A method of capturing and disposing of all this extra carbon might be found within the next couple of decades, but the additional energy involved in coal-to-liquid production means that fuel costs would inevitably be much higher than present-day prices. Worse, there is growing doubt in the energy resource sector about the long-term availability of coal. The Energy Watch Institute has suggested that a rapid increase in coal demand due to a decline in petroleum availability could even result in 'peak coal' by 2025.[4] If peak oil occurs by 2020, the chance of upscaling Australian or global coal-to-liquid production to meet rising demand seems remote. Coal liquification is likely to have insignificant impact on global petroleum prices for the foreseeable future; it will certainly offer suburban households little protection from higher petrol prices over the next several electoral cycles.

Tar sands and oil shale

Tar sands and oil shale are hydrocarbons that can be processed to produce a form of petroleum that is suitable for conversion to transport fuels. Tar sands are a gritty mix of mud, sand, rock and bitumen; oil shale is a sedimentary rock containing an organic compound called kerogen. While they have different physical properties, both can be mined and heated to separate the useful hydrocarbons into a liquid.[5] The massive amounts of energy needed to produce petroleum from tar sands and oil shale, given current technology, dramatically lessen the return on the extraction effort. Natural gas is often used to provide the needed heat – thus one 'clean' fossil fuel is burnt to make another 'dirtier' one. As natural gas prices escalate, the cost of producing petroleum from tar sands and oil shale will also rise. The environmental price of producing synthetic oil from tar sands and shale is also very high; vast volumes of water are used in processing and byproducts include large volumes of toxic waste requiring storage and eventual treatment. Even if current levels of production could be ramped up, economic and environmental constraints still signal a much higher price.

Foods to fuels

Biofuels are a favourite category of magic fuels for environmentalists. The main examples are vegetable oil, which can substitute for diesel, and ethanol, which can substitute (partly) for petrol. Biofuels are attractive because they tend to have lower carbon emissions than conventional petroleum fuels and it is assumed that they can be quickly and easily grown. This latter assumption is emerging as a big problem because global food prices have risen sharply, in part due to the increased use of crop land for biofuel production rather than food production. Between 2007 and 2008 the prices of vegetable oil (from which biodiesel is made) and grains (from which ethanol is produced) rose 97 per cent and 87 per cent respectively.[6] According to the World Bank, 65 per cent of these increases are due to increasing biofuel demand. Already the United Nations' Food and Agriculture Organisation has warned of the devastating social and economic consequences of higher food prices. Prolonged competition between biofuels and human nutrition is terrible to contemplate. Even if Australia's vehicles could be feasibly converted to run on biodiesel and ethanol, their owners will face much higher costs – either at the bowser or at the supermarket.

Plugging in

Electricity is another candidate for alternative superfuel. Electricity has long been proposed as an energy source for motor vehicles. Much of the world's public transport depends on electrically powered vehicles fed by overhead wires. But electricity is a form of energy, not a source. More than three-quarters of the world's electrical energy is generated from coal, natural gas or nuclear power.[7] Australia relies on coal for 80 per cent of its electricity,[8] so using electricity as a fuel has the same problems as other coal-based fuels. Coal-fired power stations are among the worst greenhouse gas emitters and Australia is likely to see sharp increases in electricity prices with the introduction of emissions trading.

Electric cars aren't currently produced in sufficient numbers to displace petrol-driven cars. Even the hybrid cars currently on sale only achieve fuel economy similar to high-efficiency petrol or diesel cars. Converting Australia's 11.4 million passenger vehicles to electricity would be hugely expensive. While it seems inevitable that electric cars will have wider use in the future, there is little chance of such vehicles providing similar operating convenience to current automobiles, at a similar price, at least over the medium to long term.

Hydrogen

Another pin-up überfuel candidate is hydrogen. Hydrogen is produced from water, burns cleanly and releases no greenhouse gases. Unfortunately, hydrogen doesn't exist naturally in the way fossil fuels do. Fossil fuels like natural gas must be used to produce hydrogen. Any private vehicle transport system based on hydrogen-powered vehicles risks dependence on fossil fuels, and thus problems of pollution and depletion. Also, hydrogen is a very light gas, and must be greatly compressed for storage. Hydrogen molecules are very small and, when burnt, produce less usable energy per volume than petroleum fuels. Their small size makes them prone to leakage, and hydrogen only becomes liquid under extremely high pressure, so storage canisters must be very strong and thus large and bulky. Hydrogen is also highly explosive, and there is no hydrogen supply infrastructure in Australia. While there are many hydrogen vehicle prototypes, none is nearing mainstream production. So far the only country that has seriously adopted hydrogen as an energy medium is Iceland, where abundant geothermal generation of electricity cuts fossil fuels out of the equation. It is highly unlikely that Australia's suburbs will ever run on hydrogen. They will certainly never do so at a price equivalent to petroleum.

❦

Our brief review of alternative fuels shows that there are serious doubts about their ability to replace petroleum at the same economic or environmental cost. Most of the fossil fuel alternatives use more energy in their production than fossil fuels do. Petroleum has been an extremely energy-efficient fuel to produce – every unit of energy expended on drilling and pumping oil has typically returned about 30 units of usable energy. Oil shale, tar sands and biofuels return many fewer units of usable energy than petroleum. In that sense they are really only second best. As energy constraints grow, their production costs will also increase. Coal, for example, has already rocketed in price, from around US$20 per ton in the mid 2000s to around $130 per ton in 2008. None of the so-called alternatives appears likely to reduce the future price of fuel to the levels on which suburbia was built. Another strategy is needed.

A higher plan?

The idea of planning to reduce car dependence is not new. For almost 30 years Australia's urban planners have been pursuing policies that have been, at least nominally, intended to do this. The preferred approach has been to increase the density of Australia's cities. The typical term applied to such policies is 'urban consolidation' or 'urban compaction'. Such plans have allowed a higher concentration of housing and other development within urban areas, through such measures as dual occupancy and multi-unit housing. These policies have assumed that higher urban densities will concentrate demand for public transport. Concentrated travel demand, it is presumed, will make public transport, as well as walking and cycling, more economically viable, thus reducing high energy use and greenhouse gas emissions from cars.

The view that urban density determines petroleum dependence is highly contentious in the Australian context.[9] The main scientific protagonists in this debate have been Peter Newman and Jeff Kenworthy of Curtin University, whose work has shown that per capita petrol consumption within sprawling US cities is more than twice that of Australian cities and around three times that of higher density compact European cities. (Newman has been a believer in peak oil since encountering M. King Hubbert at Stanford University in 1973.[10]) Other scientists have strongly contested Newman and Kenworthy's work; Patrick Troy has presented evidence that higher density housing may result in lower transport energy use but higher *overall* energy consumption than conventional suburban dwellings.[11] Another Australian scholar, Paul Mees, has argued that Australian cities are the same density as most US cities but their much better public transport – although poor by world standards – underpins their lower petrol consumption.[12] While their earlier work was strongly pro-densification, Newman and Kenworthy's more recent research is more nuanced. In 2006 Newman reported that access to public transport was as important as urban density in shaping household transport energy use.[13] Most recently Newman has been quoted as warning that much higher oil prices 'will mean a new residential abandonment in car-dependent suburbs. There will be wealthy eco-claves surrounded by Mad Max suburbs', and that only new rail lines (rather than higher densities) will save the suburbs.[14]

Urban consolidation policies have proven politically conten-tious, in part because the first round of consolidation allowed 'wild' densification across Australia's suburbs, especially the inner cities. The Kennett Victorian government lost office partly due to a revolt by inner-suburban preservationists angry at apartment mews sprouting next to their leafy detached villas.[15] A second, more sophisticated version of urban consolidation has tried to concen-trate higher density development around 'activity centres' or 'tran-sit-oriented developments'. Almost every Australian metropolitan strategy produced this century includes the 'concentrated activity centres' approach to urban planning.

Unfortunately, urban consolidation won't help car-depend-ent middle and outer suburbs weather the storm of higher fuel prices, especially over the medium term, because higher density urban development in Australian cities is left to private develop-ers. They respond only to housing market imperatives, which aren't yet geared to the suburban energy and environmental challenge. Australian urban housing markets are very centrally structured

– housing prices are greatest in inner zones and drop away with increasing distance from the CBD. The cheapest housing in Australian cities is generally found in the outer suburbs. The highest levels of high-density and multi-unit development are found in the inner suburbs,[16] where high land prices justify the additional cost of building up – these are the areas with the lowest petroleum dependence. There are few high-density suburban centres in Australian cities: Parramatta, Fairfield and Liverpool in Sydney are among them.

Also, the long timeframes of the development sector mean that even if a massive suburban apartment program commenced tomorrow there would be little change on the ground for many years. It would barely affect the outer suburban areas – especially those in the highest VAMPIRE categories.

Put simply, market-led urban consolidation will happen in the wrong place and at the wrong time to overcome the problem of suburban petroleum dependence (assuming that consolidation can actually reduce car reliance). This doesn't mean we should abandon the consolidation program; it means we must be more sophisticated in our use of urban consolidation as a tool for overcoming petroleum dependence. The simple relaxation of height and bulk restrictions must give way to a more nuanced but also more interventionist approach.

One of the ironies of higher fuel prices might be that density becomes less important as a determinant of public transport use. Under cheap oil conditions and without comprehensive planning, public transport found it hard to compete with the car. So planners tried to engineer higher density precincts to make walking and public transport at least more convenient. But we're now entering an era when fuel cost will be a far more important factor in travel choices: prices will be pushing people to public transport wherever they can find it.

If we still want to try to save our suburbs from the effects of car dependence through consolidation, we probably need to start placing restrictions on high-density inner-urban development to force that kind of development into suburban zones. The carbon consumption of high-density housing means we'll also need to limit the form of new compact development to the most energy-efficient scale – 2 to 4 storeys. Active state involvement in development processes will be needed to force this change; waiting for the private sector to seize the challenge will not cut it. Such planning should be combined with the deliberate relocation outwards of CBD jobs and activities, led by the public sector, to ensure that denser suburban centres are economically robust. But densification alone won't be enough to protect our oil-dependent suburban landscapes from the impact of higher fuel prices. They need a rapid transport retrofit.

Mad Max takes the bus?

Magic fuels and urban consolidation can't give Australia's cities the protection they need from higher fuel prices – another strategy is needed. Australian suburbs were founded on cheap mobility providing access to the cheap land that provided affordable housing. With this model now quivering in the face of higher fuel prices, we need to start planning for a public transport system that provides a level of mobility that can sustain our suburbs and their capacity to afford cheap housing for those on modest incomes.

A growing body of scholarly literature has shown that the quality of public transport service is a key determinant of public transport use. The better public transport is, the more likely people are to use it. Australian cities are already seeing accelerating demand for public transport due to rising fuel prices. We need now to plan new public transport services that provide a convenient travel choice for residents of the highly vulnerable VAMPIRE zones – not just for those in the inner city. Here's how it can be done.

In 2005 the European Union published its first ever guide to planning public transport networks, called *Hi-Trans*.[17] This report showed that public transport is easiest to use when it operates on clear routes with services every 10 minutes or better. When many such high-frequency routes are knitted into a well-planned network, transfers to multiple dispersed destinations become much easier. Not only do such interconnected networks speed travel times across the city; they also reduce the need to consult timetables, because the system becomes more 'legible'. This produces a kind of gestalt 'network effect', as the assembly of individual lines and routes on the public transport network link to form a much greater and more usable grid of connectivity. Such a principle is most apparent in places with compact urban form and served by metro systems, such as London or Paris. But it also applies to more dispersed cities where heavy rail systems are tightly interlinked with local bus routes, such as Toronto and Vancouver.

A key intellectual inspiration for the *Hi-Trans* public transport guide was a book by Paul Mees.[18] Mees showed, using real world examples (Toronto and Zurich), that it is possible to have effective and efficient public transport networks across a dispersed metropolitan region by linking buses, trams and trains to provide a 'go anywhere anytime' convenience and comfort similar to that of the car. The secret of these 'network effect' public transport systems is that they are rigorously and comprehensively planned by a central agency. This central agency has the power to organise and coordinate services across the network to ensure that they are tightly integrated and provide a nearly seamless travel experience. Australia's policy makers, unlike those in the European Union, do not appear to have taken much notice of Mees' work.

Centralised planning is essential to good management and coordination of public transport, but it is rare in Australian cities. In Australia, the result has been low-quality public transport with low patronage levels (nearly as low as US cities). Some Australian commentators have derided centralised public transport planning as 'Stalinist' and argue for a market-based – in reality a faith-based – system.[19] Most developed cities with good public transport, such as those in Europe, take for granted the need for a central planning authority to coordinate and integrate public transport networks. Attempts to franchise public transport services to the private sector also risk failure because the complexity of the franchising arrangements requires greater management expertise than is needed to operate the network in the first place – as the failure of Melbourne's privatised public transport has shown.[20]

Zurich has among the highest public transport patronage of any Western city. Almost obsessive levels of precision in state public transport planning have produced a system that is so good that 58 per cent of commuter travel is by public transport.[21] Trace almost any local bus line on the Zurich public transport map and it will connect with a train, tram or another bus service offering an extraordinary web of services. Zurich's transport planners even provide small dispersed low-density mountain hamlets beyond the metropolitan area with bus services at 20-minute frequencies – thus also demolishing the assumption that density is the *sine qua non* of good public transport.

Foreign examples such as Zurich, Toronto and Vancouver have already sealed the case in favour of centralised public transport planning, but Brisbane's experiment with centralised network coordination has driven the message home even more. Since 2004 the Queensland government's new Translink agency has turned a gaggle of uncoordinated bus and rail services into a system that – in parts – is starting to create the 'network effect' described by Mees. With the planning power to force the region's 18 mostly private operators to coordinate and integrate their services, Translink has seen patronage grow by 39 per cent in four years. Some of Brisbane's new high-frequency 'BUZ' bus routes have been so popular that there aren't enough buses to meet demand. The Queensland government is so pleased with its new planning authority – a popular new public brand – that it's strengthening Translink into a stand-alone statutory agency with even more power to plan and coordinate southeast Queensland's public transport network. The areas the state took over from private bus companies have shown among the greatest patronage gains. Current predictions anticipate a doubling in public transport use in southeast Queensland over the next decade.

An appreciation of the 'network' character of high-quality public transport appears implicit in some of the conceptual work underpinning urban consolidation policy. Indeed public transport planning is perhaps the 'forgotten' factor in the literature on oil-resilient urbanism. For example, Newman and Kenworthy's 1999 book showed a schematic 'network city' involving a constellation of higher density precincts linked across a web of high-quality inter-secting suburban public transport lines. This approach seems to be the inspiration for much of the fascination with activity centre and transit-oriented development (TOD) planning that is evident in recent Australian metropolitan plans. But these plans lack the coherent and systematic planning needed to deliver real 'network city' public transport. Even the activity centres and TODs are mostly located in areas where public transport is already good, not in the suburbs where oil vulnerability is greatest. This gross failure of Australia's governments and policy leaders is delaying the inevitable task of making our suburbs oil resilient.

Fortunately, good network planning doesn't necessarily require vast spending on infrastructure. Australian cities already have among the world's most extensive urban rail networks. Brisbane has one of the highest ratios of rail line to population, while Sydney's suburban rail network has more rail track than the London Underground, Paris Metro and New York Subway *combined*. Australia's rail networks suffer from extraordinarily poor management, and there is negligible integration of bus systems with the rail links. Huge productivity gains are waiting to be wrung out of Australia's public transport networks through improved network planning. For suburbs on the periphery of the rail networks, new bus routes running buses every 10 minutes or better and linked to the metro-politan rail lines could easily and dramatically enhance mobility at minimal cost. Buses are cheap – just under $500,000 for a standard model – and can be run on suburban streets without any dedicated rights-of-way. They can be given priority over cars through transit lanes and traffic light sequencing. Buses can run easily on natural gas – which Australia has in abundance – with clear greenhouse advantages and the need for only modest refuelling infrastructure compared with private cars. We need to get suburban Mad Max out of his 'pursuit special' and onto a BUZ.

Funding urban transport

A major investment program will be needed if we are to protect our suburbs from petroleum dependence. This program should focus on establishing centralised public transport planning agencies with ample funds to roll out new bus and rail services. Their aim should be to provide an even service quality across Australia's large suburban realms. While the prospect of large new spending demands is often daunting to federal and state governments, in the case of public transport there are two clear funding sources.

From roads to avoid ruin

The first set of funds for reconstructing Australia's oil-dependent suburbs with public transport can already be located: it is in federal and state governments' transport budgets. Most current transport investment is directed towards motor vehicle travel on major facilities such as tunnels, bridges, freeways and tollways. The previous federal government programmed $12.3 billion from 2004 to 2009 for major road infrastructure … but none for public transport. The new Labor government has promised to spend $20 billion over four years via Infrastructure Australia on projects that improve national productivity. There is probably no more productive use of these funds than the reconstruction of our suburban public transport systems. Although there is little indication that public transport is to be considered, the productivity requirement written into the legislation governing the funds at least invites the case for public transport to be proffered.

State governments are also complicit in the gross fiscal imbalance in favour of roads. The main transport scheme in the Victorian government's *Melbourne 2030 Metropolitan Strategy* was the

$2.7 billion Scoresby Freeway, and the government is currently contemplating a $10 billion road tunnel under Melbourne's inner city. In Brisbane, the Queensland government has supported the Brisbane City Council's $3 billion North–South Bypass Tunnel and is constructing the $1.9 billion Gateway Motorway duplication. It is also planning for the $3.4 billion Airport Link tunnel, which will compete with the Airport Rail line. If there are such vast funds available for transport it should be a relatively easy task to allocate more of them to public transport rather than roads. Such decisions are, of course, more political than technical. The experience of Western Australia has shown that eschewing the 'more and bigger roads' approach to urban transport planning can bring dividends; the success of the Southern Suburbs line and underground link is just one example. There is a desperate need for similar political leadership at state and federal level to reverse the decades-long imbalances in transport budgets.

The public transport mine

The second source of funds for urban public transport is the enormous flood of mining royalties and taxes on mining income that state and federal governments have plumbed into their treasury coffers. The minerals industry contributed $9 billion to government revenue in 2007, including $2 billion in state royalties and $5.1 billion in federal company taxes.[22] The fossil energy sector has been a big part of this massive stream of government income as the global energy hunger has stoked prices for these commodities. The oil, coal and gas sectors together sent $5.7 billion in royalties to Australian governments in 2007.[23]

The pressures that are generating these huge revenues for governments are also bringing fuel and housing pain as the rising cost of petrol pushes up interest rates. Australia's governments need to find a way of using the economic plenty from the global energy surge to assist those who are feeling that pain. In the case of transport, this means taxing Australia's fossil energy exports and redistributing the revenue to the suburbs through a massive public transport planning and investment program. This may sound politically radical, but it is already being done by the Queensland government, which kept its 2008/2009 budget in surplus by raising an additional $580 million in royalties from the state's coalmining sector. This move also seems to have also been politically popular – many Queenslanders feel entitled to a share in the mining merriment.[24] The federal government needs to start applying this principle on a national scale by using the gains from the energy boom to build energy resilience in Australia's suburbs.

The housing question

Most of our ideas for protecting Australia's suburbs from higher fuel prices focus on reconstructing suburban public transport systems. This is because transport systems support urban mobility and underpin the viability of suburban housing. Without low-cost transport, our suburbs risk becoming unviable. If our car-dependent transport systems become exorbitantly expensive, or no longer functional, our suburban housing system is in deep trouble. So any policy to transport our suburbs cheaply is also an affordable housing policy.

Form

Higher fuel prices will transform the way we plan and organise suburban housing. This doesn't just mean the design and location of suburban housing; we will also need to begin considering how to ensure that communities have adequate access to employment and services. Local-scale planning of commercial and retail activities must improve dramatically to ensure that residents of our suburbs are not forced to travel long distances to work or shop. This will mean a less permissive planning attitude to 'big box' drive-in retail complexes and a focus on more smaller scale walkable 'high street' strips around the nodes of the new high-frequency public transport links.

The shift to higher residential densities will inevitably continue apace in our cities. Far more attention needs to be given to ensuring that this intensification is distributed and coordinated with retail and commercial activities, public transport and amenities such as public open space. This kind of planning is largely absent in contemporary consolidation policy. Urban planning must also abandon its obsession with inner-city high-rise apartments in favour of concentrations of 2 to 4-storey attached dwellings around key public transport nodes. This reconstruction by planning will require greater public control of private development processes, including strengthening urban growth boundaries to contain car-dependent outward expansion, much tighter precinct planning at the local scale and the upskilling of local government planners (and politicians) to ensure that they are capable of these new tasks. Unfortunately, changing the built form of our cities is a slow process – we will be stuck with our current built environment far into the future.

Tenure

We need to begin rethinking our heavy dependence upon owner-occupier suburban housing. Owner occupation depends on cheap mortgage finance, and it is already clear that the era of cheap mortgages is over and that the rising cost of transport for distant and dispersed suburban lots will enter into the calculations of mortgage lenders. Many suburban households will be unable to afford housing finance. We must begin thinking of alternative tenure forms for those who seek long-term security but cannot get financing to purchase a dwelling. The private rental sector does not offer this security, as length of tenure is largely at the whim of landlords.

We don't have space in this book to make a case for restructuring of Australia's housing tenure system. But if we are serious about reconstructing our suburbs, we will need to do just that. There are many models available. There is the northern European 'cost renting', in which rents are set relative to the costs of construction rather than on the basis of investment and profit. Another model might be mass social housing. Australia's greatest urban economist, Hugh Stretton, has proposed a system of non-market home owning in which governments would build and sell affordable housing to modest-income households through low, fixed-rate mortgages with a contractually fixed rate of capital gain, subject to the requirement that the property be 'sold' only to the government and never enter the private market.[25] Also, collective and quasi-collective suburban housing solutions, currently only niche operations, can operate at larger scales. Mad Max may well find that a 3-storey walk-up apartment a few steps from a cross-suburban BUZ offers a convenient alternative to outback barbarism.

Network planning

Reconstructing our cities and suburbs for the era of higher fuel prices must begin immediately. We can't rely on fantasy fuels and magic machines to save our suburbs. We desperately need national planning to mitigate the effects of the VAMPIRE on our suburbs. This implies strong and far-sighted political leadership. At the time of writing there were some glimpses of this with Prime Minister Kevin Rudd opining that people should catch the bus to offset the impact of fuel prices on household budgets. But this is only viable for those with services that link their area with metropolitan networks. We need courageous leadership to translate this idea into a comprehensive plan.

Our current federal government and its state counterparts must respond to higher fuel prices if they are to achieve electoral longevity. They will need to abandon some of the mantras of contemporary Australian transport and energy policy. First, any hope for wondrous technology or fuel must be replaced by the tried and tested technology of buses, trains and trams. Next, this old technology must be accompanied by strong central planning rather than weak, disparate and uncoordinated market forces. This also means less emphasis on urban consolidation and more on network planning. Any new funds required for suburban reconstruction can be taken from existing budgets or mined from the vast open-cut pits of energy export revenue. The energy crisis may be long and challenging, but with good planning and sound leadership it need not traumatise our suburbs.

CHAPTER 7

Aftershocks: Facing the challenge of oil vulnerability in Australia's cities

We started assessing the oil vulnerability of Australia's cities when the price of oil was US$40 per barrel. We began writing this book when the price of oil was around $60 per barrel. By the time we handed the manuscript to our publisher, oil prices sat at $134 per barrel. We've found it hard to keep up with the weekly changes in the oil price. In just a few months in 2008 we've seen fuel prices move from the fringe of public consciousness to the centre. As we sent in our manuscript, emergency meetings of the world's oil-consuming nations – including Australia – and global petroleum suppliers, dominated by OPEC were being thrown together. Such deliberations will only intensify as fuel prices march ever higher.

The early months of 2008 have offered us a clear glimpse of a strange and unfamiliar world, and the coming months and years seem likely to be even stranger. Uncertainty is the only certainty with global petroleum futures. It's impossible to tell what the trajectory of oil prices will be over the next few years, what their rate of climb – or collapse – will be. One thing is almost certain: if the global desire for oil continues to increase, the price is unlikely to fall. Projections of the growth and development of Australia's cities must now assume higher fuel prices. This will jolt our cities and suburbs.

The price of private transport will soar far beyond the level on which our suburbs were planned. Our cities will face their greatest transformation since the postwar surge of suburbanisation. If the astonishing success of Australia's suburbs as the heartlands of a settled and decent national life is to continue, we must act now. Those on the lowest incomes in the most car-dependent outer suburbs will face the greatest burden in this new world.

The impacts of higher fuel prices and interest rates take time to percolate through our urban economies: today's petrol pain and mortgage anxiety are the result of pressure over the last few years, so the full impacts of today's high fuel costs won't be felt for several years. We should be planning now for those impacts.

We have shown that one impact will be a reduced ability, financially speaking, to travel around our metropolises. The added pressure of inflation and interest rates could rupture the basic land, transport and financial relations that underpin suburbia. At its most basic level, suburbanisation is a method of providing housing for the masses through cheap peripheral land in otherwise costly metropolitan housing markets. Credit for such housing, in part, presumes that land values will be sustained, which in turn presumes cheap transport. In the early decades of suburbanisation, access to the suburbs was provided by the rail networks. After World War II, the means of access shifted to the car. Higher oil prices could fracture that housing and transport nexus, and that might spark the flight of people and capital back from the suburban frontier. The dream of suburban home ownership could shatter irreparably. Already one of Australia's most senior urban thinkers has warned that residents may abandon the urban perimeter, and has drawn on Australia's cinematic imagination to invoke a vision of a post-petroleum wasteland.

The complete functional collapse of Australian suburbia under the strain of higher fuel prices is plausible, but improbable. Nonetheless it will take dedicated effort to maintain the quality and practicality of suburban living over the coming decades. Australia's cities (and citizens) need a new plan, a plan in which there is public transport access for all. Those in the outer tracts must be given

equal standing with respect to public transport services and infrastructure, just as they should be given equal standing with public
infrastructure such as schools, hospitals, and emergency services.

Comprehensive planning by Australian governments will be
needed. This must start immediately – it is the only sure means of
protecting our cities and their residents from the looming energy
shocks. It must include building capacity within our governments
to engage with and intervene in our suburbs. It must not assume
that hypothetical markets can produce security, or blindly put its
faith in technology to resolve our transport woes.

Only government can provide the detailed and comprehensive planning needed for the task of spreading transport equity
across Australia's cities. Only government can assemble the funds
needed for this new scale of infrastructure and services provision.
Only government agencies can plan and coordinate public transport networks to provide a 'go anywhere anytime' service to all our
urban residents. This planning must begin now so that the age of
insecure oil does not end up shocking our suburbs.

Acknowledgments

The authors would like to thank colleagues within Griffith University's Urban Research Program and School of Environment, and those beyond, for their insights and observations on the conditions of Australia's cities and for their interest in our work. We are especially grateful to Brendan Gleeson for allowing us to work in what is Australia's most engaged and energetic urban research centre. Rick Evans of the Urban Research Program deserves particular thanks for his technical expertise in assisting with the VAMPIRE index. Beyond the academy, Catherine Townsend is the recipient of further gratitude for her keen and constructive feedback on the drafts, which did nothing but improve the final copy.

Responsibility for all errors and omissions contained herein lies, of course, with the authors.

Notes

2 Solving housing problems by creating suburbs

1   Freeland, J.M. (1972), 'People in Cities', in A. Rapoport (ed.), *Australia as Human Settlement: Approaches to the Designed Environment*, Angus & Robertson, p. 119.
2   Wohl, A. (1977), *The Eternal Slum: Housing and Social Policy in Victorian London*, Edward Arnold.
3   Stretton, H. (1975), *Ideas for Australian Cities*, Georgian Press; Mullins, P. (1981), 'Theoretical Perspectives on Australian Urbanisation: 1. Material Components in the Reproduction of Labour Power', *Australian and New Zealand Journal of Sociology* 17(1), p. 65.
4   Davison, G. (1978), *The Rise and Fall of Marvellous Melbourne*, Melbourne University Press.
5   Kass, T. (1987), 'Cheaper than Rent: Aspects of the Growth of Owner-Occupation in Sydney 1911–1966', in M. Kelly (ed.), *Sydney: City of Suburbs*, UNSW Press, pp. 77–94.
6   Berry, M. (1988), 'To Buy or Rent?: The Demise of a Dual Tenure Policy', in *New Houses for Old: Fifty Years of Public Housing in Victoria, 1938–1988*, Victorian Ministry of Housing and Construction.
7   Jones, M.A. (1972), *Housing and Poverty in Australia*, Melbourne University Press, p. 140.
8   Hill, M.R. (1959), *Housing Finance in Australia*, 1945–1956, Melbourne University Press.
9   Badcock, B. (2000), 'Home Ownership and the Illusion of Egalitarianism', in P. Troy (ed.), *A History of European Housing in Australia*, Cambridge University Press.
10   Neutze, M. (1977), *Urban Development in Australia*, George Allen & Unwin.
11   Allport, G. (1987), 'Castles of Security: The New South Wales Housing Commission and Home Ownership 1941–1961', in Kelly, *Sydney: City of Suburbs*, pp. 95–124.
12   Neutze, M. (1978), *Australian Urban Policy*, George Allen & Unwin.
13   Neutze, *Urban Development in Australia*.
14   Randolph, B. (2002), 'Third City Suburbs: Options for housing policy in ageing middle ring suburbs', *Australian Planner* 39, pp. 173–78.
15   Queensland Transport (2005), *Smart Travel Choices for South East Queensland: A Transport Green Paper*, Queensland Government.

16 Western Australian Planning Commission (2004), *Network City: Community Planning Strategy for Perth and Peel*, WA Government.
17 Dodson, J. and Sipe, N. (2007), 'Oil Vulnerability and Urban Planning', *Planning News* 33(8), pp. 12–13.
18 Department of Infrastructure, Planning and Natural Resources (2003), *Regional Transport Indicators for Sydney*, NSW Government.

3   Towards uneasy oil

1 Chevron Oil Ltd (2005), advertisement in *The Economist*, 16 July, pp. 6–7.
2 Worldwatch Institute (2007), *Renewables 2007: Global Status Report*, Worldwatch Institute.
3 United Nations (2007), *World Population Prospects: The 2006 Revision*, Department of Social and Economic Affairs, United Nations Population Division.
4 International Monetary Fund (IMF) (2007), 'World Economic Outlook Database, April 2007', <http://www.imf.org/external/pubs/ft/weo/2007/01/data/download.aspx>.
5 Ibid.
6 Yi-chong, X. (2007), 'China's Energy Security', in M. Wesley (ed.), *Energy Security in Asia*, Routledge.
7 Ibid.
8 International Energy Agency (IEA) (2008), *Oil Market Report: March*, IEA and Organisation for Economic Cooperation and Development (OECD).
9 Simmons, M. (2005), *Twilight in the Desert: The Coming Saudi Oil Shock and the World Economy*, John Wiley & Sons.
10 LeBlond, D. (2005), 'Investment cited as most likely limit to oil supply', *Oil and Gas Journal*, 18 July, p. 25.
11 IEA (2006), *World Energy Outlook 2006*, IEA and OECD, p. 40.
12 Izundu, U. (2008), 'OTC speakers highlight offshore industry's future', *Oil and Gas Journal*, 12 May, p. 20.
13 Klare, M. (2005), *Blood and Oil: The Dangers and Consequences of America's Growing Petroleum Dependency*, Penguin.
14 Energy Information Administration (2005), *World Oil Transit Chokepoints*, US Department of Energy.
15 Deffeyes, K. (2001), *Hubbert's Peak: The Impending World Oil Shortage*, Princeton University Press.
16 Hubbert, M.K. (1956), *Nuclear Energy and Fossil Fuels*, Proceedings of American Petroleum Institute Drilling and Production Practice, 7–9 March, American Petroleum Institute.
17 Australian Petroleum Production and Exploration Association (2007), 'Australian Petroleum Production Statistics (Microsoft Excel Spreadsheet)', retrieved 31 March 2007 from <http://www.appea.com.au/Statistics/documents/AnnualProductionStatistics.xls>.
18 Geoscience Australia (2006), *Oil and Gas Resources of Australia 2004*, Geoscience Australia.
19 Bentley, R.W. (2002), 'Global Oil & Gas Depletion: An Overview', *Energy*

Policy 30, pp. 189–205; Hirsch, R. (2007), *Peaking of World Oil Production: Recent Forecasts*, Department of Energy and National Energy Technology Laboratory.

20   Deffeyes, *Hubbert's Peak: The Impending World Oil Shortage.*

21   Bakhtiari, A.M.S. (2004), 'World Oil Production Capacity Model Suggests Output Peak by 2006–07', *Oil and Gas Journal*, 26 April, p. 18; Zittel, W. and Schindler, J. (2007), *Crude Oil: The Supply Outlook*, Energy Watch Group.

22   Yergin, D. (2006), *Why the 'Peak Oil' Theory Falls Down: Myths, Legends, and the Future of Oil Resources*, Cambridge Energy Research Associates.

23   Lynch, M.C. (2003), 'Petroleum Resources Pessimism Debunked in Hubbert Model and Hubbert Modelers' Assessment', *Oil and Gas Journal*, 14 July, p. 38.

24   Hirsch, *Peaking of World Oil Production: Recent Forecasts.*

25   Heinberg, R. (2003), *The Party's Over: Oil, War and the Fate of Industrial Societies*, New Society Publishers; Campbell, C. (2005), *Oil Crisis*, Multi-Science Publishing.

26   Huber, P. and Mills, M.P. (2004), *The Bottomless Well: The Twilight of Fuel, the Virtue of Waste, and Why We Will Never Run Out of Energy*, Basic Books.

27   Heinberg, *The Party's Over: Oil, War and the Fate of Industrial Societies*; Kunstler, J.H. (2005), *The Long Emergency: Surviving the Converging Catastrophes of the Twenty-first century*, Grove/Atlantic; Simmons, *Twilight in the Desert: The Coming Saudi Oil Shock and the World Economy.*

28   Deffeyes, *Hubbert's Peak: The Impending World Oil Shortage*; Deffeyes, K. (2005), *Beyond Oil: The View from Hubbert's Peak*, Farrar, Straus, and Giroux.

29   Bentley, 'Global Oil & Gas Depletion: An Overview', pp. 189–205.

30   Greene, D.L., Hopson, J.L. et al. (2006), 'Have We Run Out of Oil Yet? Oil Peaking Analysis From an Optimist's Perspective', *Energy Policy* 30, pp. 515–31.

31   Odell, P. (2004), *Why Carbon Fuels Will Dominate the 21st century's Global Energy Economy*, Multi-Science Publishing.

32   IEA (2004), *World Energy Outlook 2004*, IEA and OECD.

33   Ibid.

34   IEA, *World Energy Outlook 2006.*

35   IEA (2007), *World Energy Outlook 2007: Focus on China and India*, IEA and OECD.

36   Birol, F. (2008), 'We Can't Cling to Crude: We should leave oil before it leaves us,' *Independent*, retrieved 19 March 2008 from <http://www.indpendent.co.uk>.

37   Commission on Oil Independence (2006), *Making Sweden an Oil-Free Society*, Government of Sweden.

38   Hirsch, R., Bezdek, R. et al. (2005), *The Peaking of World Oil Production: Impacts, Mitigation and Risk Management*, Department of Energy, Washington.

39   Fournier, D.F. and Westervelt, E.T. (2005), *Energy Trends and Their Implications for U.S. Army Installations*, US Army Corps of Engineers.

40   Volvo (2008), *Future Fuels for Commercial Vehicles*, AB Volvo.

41   Chevron Oil Ltd (2005), advertisement in *The Economist*, 16 July, pp. 6–7.

42   Quoted in Bergin, T. (2006), 'Total Sees Output Peak, Urges Less Demand', Reuters News.

43 van der Veer, J. (2008), 'Two Energy Futures', retrieved 18 March 2008 from <http://www.shell.com/home/content/aboutshell-en/our_strategy/shell_global_scenarios/two_energy_futures/two_energy_futures_25012008.html>.

44 Department of the Prime Minister and Cabinet (2004), *Securing Australia's Energy Future* Australian Government, p. 119.

45 Ferguson, M., Minister for Resources and Energy (2008), speech to Third Annual Coal-to-Liquids and Gas-to-Liquids Conference, Brisbane, 18 February.

46 Senate Standing Committee on Rural and Regional Affairs and Transport (2007), *Australia's Future Oil Supply and Alternative Transport Fuels: Final Report*, Parliament of Australia, p. 55.

47 Queensland Oil Vulnerability Taskforce (2008), *Queensland's Vulnerability to Rising Oil Prices*, Queensland Government.

48 Brisbane City Council and Maunsell Australia (2007), *Climate Change and Energy Taskforce Final Report*, Brisbane City Council.

4 Shocking the city: Higher fuel and housing costs in the suburbs

1 Walliker, A., Ife, H. and Cogdon, K. (2006), 'Families Feel Pinch As Bills Mount Up', *Herald Sun*, 4 May, p. 5.

2 <http://www.aip.com.au/pricing/retail/monthly/index.htm>.

3 Australian Bureau of Statistics (ABS) (2005), *Household Expenditure Survey 2003–2004*.

4 National Roads and Motorists' Association Limited (NRMA) (2007), *Private Whole of Vehicle Operating Costs – June 2007*, NRMA Motoring and Services.

5 *Daily Telegraph* (2006), 'Tightening the Mortgage Belt', 6 May, p. 62.

6 Translink (2007), Translink Draft Network Plan, Queensland Transport.

7 Heger, U. (2008), 'Council's Bus Squeeze Crisis', *Courier Mail*, 12 May.

8 Department of Infrastructure (2008), *East West Link Needs Assessment Final Report*, Victorian Government, p. 72.

9 RailCorp (2006), *Annual Report 2005–2006*, RailCorp NSW, p. 6; RailCorp (2007) *Annual Report* 2006–2007, RailCorp NSW, p. 10.

10 Wade, M. (2006), 'Guzzlers' Value Slips As Fuel Costs Rise', *Sydney Morning Herald*, 22 July, p. 7.

11 McKinley, J. (2008), 'A Run on Rice in Asian Communities', *New York Times*, 1 May.

12 Arends, B. (2008), 'Load Up the Pantry', *Wall Street Journal*, 21 April.

13 Authors' calculations using REIA and Mortgage Choice (2008), 'Market Facts: Quarterly Moving Annual (Trend) Median House Prices' (spreadsheet).

14 Reserve Bank of Australia (RBA) (2008), 'Household Finances: Selected Ratios' (spreadsheet), RBA.

15 Johnson, B., Manning, S., Disney-Willis, D. and North, M. (2008), *Australian Mortgage Industry: Volume 7*, Fujitsu Consulting and JP Morgan.

16 Saulwick, J. (2008), 'Huge Rise in Home Evictions', *Sydney Morning Herald*, Sydney, 23 January, p. 1.

17 Dart, J. and Irvine, J. (2008), 'When Pain Persists, They Come Knocking',

Sydney Morning Herald, Sydney, 28 March, p. 1.

18   Walsh, K. and Singer, M. (2006), 'Left With an Empty Feeling', *Sydney Morning Herald*, 10 September, p. 5.

19   Saulwick, J. (2007), 'How the West Was Lost', *Sydney Morning Herald*, 8 December, p. 29.

20   Dodson, J. and Sipe, N. (2006), *Shocking the Suburbs: Urban Location, Housing Debt and Oil Vulnerability in the Australian City*, Research Paper 8, Urban Research Program, Griffith University.

21   See Dodson, J. and Sipe, N. (2008), *Unsettling Suburbia: The New Landscape of Oil and Mortgage Vulnerability in Australian Cities*, Research Paper 17, Urban Research Program, Griffith University, <http://www.griffith.edu.au/_data/assets/pdf_file/0003/88851/urp_rp17_dodsonsipe-2008.pdf> (from November 2008).

5 Oil and politics: The impact of rising fuel costs on suburban constituencies

1   Pearce, C. (2002), Address In Reply, 14 February, <http://www.chrispearcemp.com/speeches.php?article=422>.

2   Hawley, J. (2003), 'Crowded Land of Giants', *Sydney Morning Herald, Good Weekend*, 26 August.

3   Wilson, S. and Turnbull, N. (2000), 'Howard's Secret Keynesiansim', *Australian Review of Public Affairs*, 11 August.

4   Australian Institute of Petroleum (2008), 'National Metropolitan Daily Retail Unleaded Petrol Price', retrieved 13 May 2008 from <http://www.aip.com.au/pricing/retail/monthly/index.htm>.

5   Howard, J. (2005), interview with Tanya Nolan on ABC Radio, *The World Today*, 26 September.

6   Tingle, L. (2006), 'Put a Tax Cut in Your Tank', *Australian Financial Review*, 11 May, p. 1.

7   Rudd, K. (2007), 'Do You Feel Like You Have "Never Been Better Off"?', Australian Labor Party advertisement, <http://www.youtube.com/watch?v=0dU02l8savA&mode=user&search=>.

8   Australian Bureau of Agricultural and Resource Economics (ABARE) (2004–07), *Australian Commodities*, December issues.

9   Dodson, J. and Sipe, N. (2008), *VAMPIRE at the Polling Booth: Oil and Mortgage Vulnerability and the 2007 Australian Federal Election*, Research Paper 19, Urban Research Program, Griffith University.

10   Australian Competition and Consumer Commission (ACCC) (2007), *Petrol Prices and Australian Consumers: Report of the ACCC Inquiry into the Price of Unleaded Petrol*, ACCC.

11   Australian Government (2008), *Budget Papers: 2008–2009*, np, Table 6.

12   Currie, G. and Senbergs, Z. (2007), *Exploring Forced Car Ownership in Metropolitan Melbourne*, Proceedings of the 30th Australasian Transportation Research Forum, 25–27 September.

13   Transport and Population Data Centre (2006), *Household Travel by SLA in Sydney*, NSW Government.

14 Coorey, P. (2006), 'MP Ditches Car for Public Transport', *Sydney Morning Herald*, 18 April, p. 3.
15 Dodson, J. and Sipe, N. (2008), 'Planned Household Risk: Mortgage and Oil Vulnerability in Australian Cities', *Australian Planner* 45(1), pp. 38–47.
16 Mees, P. (2005), 'Privatization of Rail and Tram Services in Melbourne: What Went Wrong?', *Transport Reviews* 24(4), pp. 433–49.

6 Oil-proofing Australian cities: Transport and planning solutions to oil vulnerability

1   Fisher, Brian (2006), quoted in 'Peak Oil', *Four Corners*, ABC Television, 10 July, <http://www.abc.net.au/4corners/content/2006/s1683060.htm>.
2   Ibid.
3   Darley, J. (2004), *High Noon for Natural Gas: The New Energy Crisis*, Chelsea Green Publishing.
4   Energy Watch Group (2007), *Coal: Resources and Future Production*, Energy Watch Group.
5   Deffeyes, *Beyond Oil: The View from Hubbert's Peak*.
6   Food and Agriculture Organisation (FAO) (2008), *Soaring Food Prices: Facts, Perspectives, Impacts and Actions Required*, background paper for High Level Conference on World Food Security: The Challenges of Climate Change and Bioenergy, Rome, 3–5 June, United Nations.
7   IEA, *World Energy Outlook 2007: Focus on China and India*.
8   Ferguson, M., Minister for Resources and Energy (2008), 'Budget Boosts Clean Coal and Renewable Energy', Media Release, 13 May.
9   Troy, P. (1996), *The Perils of Urban Consolidation: A Discussion of Australian Housing and Urban Development Policies*, Federation Press; Searle, G. (2004), 'The Limits to Urban Consolidation', *Australian Planner* 41(1), pp. 42–48; Gray, R. and Gleeson, B. (2007), *Energy Demands of Urban Living: What Role for Planning?*, 3rd National Conference on the State of Australian Cities, 28–30 November, University of South Australia and Adelaide University; Rickwood, P., Glazebrook, G. et al. (2008), 'Urban Structure and Energy: A Review', *Urban Policy and Research* 26(1), pp. 57–81.
10 Newman, P. (2006), 'After Peak Oil: Will Our Cities and Regions Collapse?', submission to Australian Senate Inquiry into Australia's Future Oil Supply and Alternative Transport Fuels, Senate Rural and Regional Affairs and Transport Committee.
11 Troy, *The Perils of Urban Consolidation: A Discussion of Australian Housing and Urban Development Policies*.
12 Mees, P. (2000), *A Very Public Solution: Transport in the dispersed city*, Melbourne University Press.
13 Newman, P. (2006), 'Transport Greenhouse Gas and Australian Suburbs: What Planners Can Do', *Australian Planner* 43(2), pp. 6–7.
14 Newman, P. cited in Campion, V. (2008), 'Transport Crisis Turns West into Wasteland', *Daily Telegraph*, 1 May.
15 Lewis, M. (1999), *Suburban Backlash: The Battle for the World's Most Liveable City*, Bloomings Books.
16 Randolph, B. (2006), 'Delivering the compact city in Australia: Current trends

and future implications', *Urban Policy and Research* 24(3).

17   Nielsen, G. and Lange, T. (2005), *Hi-Trans: Planning the Networks*, Stavanger (Norway), European Union Interreg III and Hi-Trans.

18   Mees, *A Very Public Solution: Transport in the dispersed city*.

19   Moran, A. (2008), 'Melbourne's public transport is on a road to nowhere', *The Age*, 7 April, p. 8.

20   Kain, P. (2007), 'The pitfalls in competitive tendering: Addressing the risks revealed by experience in Australia and Britain', staff paper, Bureau of Transport and Regional Economics, Canberra.

21   Zurcher Verkehrsverbund (2008), Marktdaten (market data), retrieved 4 June 2008 from <http://www.zvv.ch/holding_kennzahlen5.asp>.

22   Minerals Council of Australia (2008), 2008–2009 Pre-Budget Submission, Minerals Council of Australia, p. 4.

23   Department of Foreign Affairs and Trade (DFAT) (2008), *About Australia: Resources Sector*, Australia Fact Sheet Series, DFAT.

24   *Courier-Mail* (2008), 'Safe sailing if economic climate stays fine' (Editorial), 4 June.

25   Stretton, H. (2005), *Australia Fair*, UNSW Press.